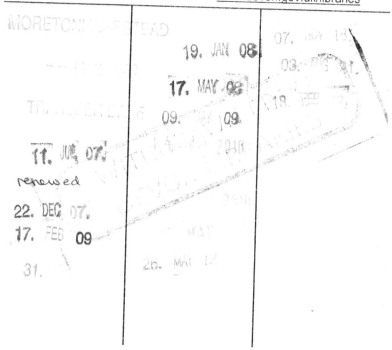

For my Dad, Bill Smaldon (1907 - 1979), who
first interested me in the natural world, and for my
Grandpa, George Morcom (1879 - 1964), who
taught me of birds - and first showed me Dartmoor.

The Birds of Dartmoor

Roger Smaldon

With Illustrations by Mike Langman

Isabelline Books

TABLE OF CONTENTS

Isabelline Books
8 Woodlane Crescent
Falmouth
Cornwall
TR11 4QS

ISBN 0-9542955-8-7 Hardback (edition limited to 100 copies)
ISBN 0-9542955-7-9 Paperback

ILLUSTRATIONS

MAPS
(following page 236)

ACKNOWLEDGEMENTS AND THANKS

No one person writes a book such as this. You may be putting the words on the paper, but you are constantly aware that you are distilling the observations of many, many people, who over the decades have put in much time, and hopefully gained much enjoyment, from studying the birds of Dartmoor.

The bulk of the records used to complete this book have come from two main sources. First and foremost, there are the records of the Devon Bird Watching and Preservation Society. The observations published in the *Annual Reports* since the Society's conception in 1929, have formed the backbone of this publication, and I must thank the Officers and Council of the Society for allowing me free use of the published records. Likewise articles published in the Society's journal *Devon Birds* have been of great help. The second source has been the observations of members of the Dartmoor Study Group. This Group was formed in 1991 to study the birdlife of Dartmoor, and improve our knowledge of species throughout the year. Its *Newsletters* since 1991, and *Bird Reports* since 1996 have been used extensively. The Group has relied totally on the enthusiasm of a relatively small number of members, many of whom have been contributing since its beginning or shortly thereafter. So if not mentioned elsewhere for specific reasons, grateful thanks to Bob Barrow, Bill Beatty, Jim Braven, Jeremy Chesher, Roy Croker, Gary and Anna Easton, Dave Glaves, George and Julia Harris, Ann Hughes, Sue Leythorne, Rob Lillicrap, Michael Mitchell, Mike Trott, Nick Walter, and John Walters.

Special thanks to Richard Hibbert, Editor of the *Dartmoor Bird Report*, for his interest in this publication, and his great help in keeping me up to date with recent records.

Mike Langman also deserves special thanks for providing superb illustrations and a really excellent painting for the frontispiece and cover. Mike's help and co-operation was always greatly appreciated.

Anyone who attempts to write on Dartmoor's birds owes a huge debt of gratitude to Dr. Peter Dare and the late Ivor Hamilton. Their work on the birds of the Postbridge area during the years 1956 to 1967, with Peter's update and comparisons in the early 1990s, was of a scale and complexity that had never been attempted before, or bettered since. Thanks also to Peter for his additional details of certain records during the early period that put meat on the bones of previously published records, notably the Fernworthy Long-tailed Skua.

Certain references have been a little difficult to obtain, and great help was obtained in this regard from Roger Swinfen and John Randall. Grateful thanks to them both, and thanks also to Dave Jenks for so willingly making available his expertise, and information gained over the years whilst researching his *A History of Devonshire Ornithology*.

Norman Baldock, Ecologist with the Dartmoor National Park Authority, generously gave permission for the Dartmoor Moorland Breeding Bird Survey 2000

Report, to be used, and also was very helpful in pointing me in the right direction regarding maps.

Many people were very helpful in providing additional information and assistance on certain species, either directly to me for the book or previously by contributing material for Dartmoor Bird Reports. Amongst these must be mentioned Mike Sampson, John St. Leger and Bob Jones for, between them, providing practically all that has been found out in recent years on the abundance, numbers and locations of Dunlin and Golden Plover on the moor. Mark Darlaston and Robin Khan were very helpful regarding raptors, and clarifying certain points on the more sensitive species. Tony John was his usual helpful self when it came to questions on Dippers, and Dave Bubear assisted greatly when it came to additional records for the Prewley/Okehampton area, especially the breeding Red-backed Shrikes at Meldon in the 1970s. Harvey Kendall was kind enough to check old records regarding Red-backed Shrikes in the Tavistock area, and to supply a hitherto unknown record of breeding Wrynecks. Dave Jenks made available all relevant information from past surveys of Rooks in the area. Grateful thanks to all, and also to Glyn Avery, Stan Christophers, Ray Gould, Ted Griffiths, Paul and Barbara Heatley, June Smalley, and Vic Tucker for providing additional records and useful comments and suggestions. Thanks also, regarding additional records, to David Ballance who very kindly gave me the details of an early Merlin's nest from W. Walmsley White's *Ornithological Journal* (Isabelline 2004).

Devon in general has never been known for its big involvement in long running BTO surveys. With this in mind, it is therefore truly fortunate that one of Devon's handful of Common Bird Census sites should fall within our area. Rob Hubble and Stella Tracey have put in an enormous amount of work over the years of the Census at Harford, and their results have proved of great value to me. I would also like to thank Rob for his work on the River Plym Water-ways Bird Survey, the full stretch covered being within the National Park, and also Neil Trout for covering a similar stretch on the River Meavy. The published results of both these surveys have been most helpful. Also David Price should receive acknowledgement for similar work put in over a long period at Dunsford Wood Nature Reserve. The type of long term work involved with these surveys is extremely important, and their value can not be over estimated. Also here a note of thanks should be given to Sarah McMahon, Rob Lillicrap and Martin Dawe for the national survey work and general observations carried out over many years in the china clay areas of Portworthy Dam, Lee Moor and Crownhill Down.

Three people, with long associations with Dartmoor, have contibuted past memories through the Dartmoor Study Group, and these have been incorporated into the general story. Lawrence Slade, moorland walker and birdwatcher since the 1940s, has recollected much of his early searches for Dunlin and Golden Plover on the northern moor, together with days watching Black Grouse and Montagu's Harriers - both now long gone. Richard Waller has told of the early days at his family home at Thornworthy, and the differences in bird life there between then and now, and

Sylvia Needham has contributed much on the changes that have occurred in the wild life in the area of her farms, firstly at Throwleigh and then for a great many years at Widecombe. Background information like this is truly invaluable, when attempting a study of birds in a given area.

When combing through bird reports from years ago, one is constantly reminded that although many people have submitted records, a comparative few contributed the bulk of the sightings. Their names appear again and again, reminding us of their long-standing interest in the region, and the great enjoyment that they must have gained from the hours spent in this very special landscape. Several of these people have been mentioned already. Some are happily still contributing to our knowledge of the moor, but unfortunately many from the earlier years are no longer with us to enjoy Dartmoor's wild country. They have all helped in building our knowledge of the moor's birds, and should receive thanks. Thanks therefore firstly to H.G. Hurrell and Malcolm Spooner, who between them put so much Dartmoor material into the early *Devon Bird Reports*, to Douglas Gordon, W. Walmesley White, F. Howard Lancum, E.H. Ware, Pamela Lind, Roy Curber, and Ray Smith. To Lawrence Slade and Tony Archer-Lock, John Wesley and Charles Sawle, Leonard Hurrell, Peter Goodfellow, Gordon Vaughan, and Geoff Weymouth. Apologies to any Dartmoor devotees past or present who, through my oversight, have failed to get a mention.

Thanks to Michael Whetman of Isabelline Books for help and guidance, and for first approaching me on the matter of a book that I had thought about for many years, but somehow had never quite started. His approach gave me the impetus I needed.

Needless to say, any mistakes or omissions are mine alone. There will inevitably be some, and I would ask that if found, they could be made available to me via the publisher. This would greatly assist the correction of any future edition.

Finally a big thank you to my family, to my brother Geoff for help and professional advice, to my son Matthew who assisted with getting much of the manuscript computerised in the early days, to my other son Adam, and especially to my wife Aileen for putting up with all the hours I spent, of necessity, lost deep in research or getting sections finalised.

INTRODUCTION

Dartmoor is unique. Many people with totally different interests in the moor find this, the largest upland area in southern England, exceptional for various reasons. For anyone interested in birds Dartmoor is special because as well as the obvious attractions of an upland area and the changes in the general avifauna from surrounding lowland, there are certain things that set it apart from anywhere else in Britain.

As an upland region Dartmoor is ornithologically something of a watershed. Two hundred and fifty-four species and sub-species species have been recorded there to date. Its hills and coombes are home to certain breeding species that extend no further south in the world, and it also holds populations, albeit small ones, of species that are on the northern limit of their range. Although not holding the sheer numbers of birds found on more mountainous uplands to the north, it does have a mixture of lowland heath and upland species that is unique in these islands. Breeding Golden Plover and Dunlin are at their southerly limit on Dartmoor hills. No others breed further south in the world. The Red Grouse population is the most southerly in Europe, although it is from stock introduced long ago. Yet the Red Grouse are far more likely to have the heathland Hobby hunting overhead in the summer than the Merlin of more northern regions. Another heathland species, the Dartford Warbler, is spreading higher into suitable gorse habitat every year, and Dartmoor is the only place on earth where Dartford Warblers and Dunlin and Golden Plover breed within a few miles of each other. Other than possibly an area of Brittany, Dartmoor is the only place in Europe where it is just possible to hear the northern Ring Ouzel, of the nominate race *torquatus,* and the southern Dartford Warbler singing within hearing distance at the same site.

The area covered in this book is basically the area of Dartmoor National Park. Dartmoor was designated a National Park in 1951. The area covers 368 square miles (953 square kilometres). One of the joys of Dartmoor is that there is so many different types of habitat within a relatively small area. Most of the Park is unenclosed open moorland, but there are many deciduous woodland areas in the river valleys, conifer plantations, upland bog, valley mire, heathland downs and commons on the perimeter, and a considerable area of agricultural land, especially on the eastern borders of the Park. Not surprisingly most records in the book will refer to the wilder upland areas, the woodlands and plantations. Every effort has been made to include records of farmland birds from the border regions of the Park, although these have never received the attention given to the true moorland birds. There are certain, mostly small, areas of moorland in the area, that are outside the National Park boundaries. The most important of these were excluded from the National Park on its designation for political reasons. As they are part of the natural moorland area they are included in this book. The primary area is the china clay region around Lee Moor, including Shaugh Moor, Crownhill Down, and the ornithologically important area of pools and settling pits at Portworthy Dam. Other notable areas are private marshland just off the moor near Sourton, and the extreme western area of Roborough Down above Grenofen, and the nearby West Down.

Many publications give details of Dartmoor habitats. One of the most useful and recent is *The Nature of Dartmoor : A Biodiversity Profile* published by the Dartmoor National Park Authority and English Nature in 2001, and obtainable from the DNPA shops. Much of the following brief sketch of habitat, with particular relevance to birds, is based on that profile.

Blanket Bog

This, the most southerly blanket bog in Britain and almost in Europe, covers about 120 square kilometres of Dartmoor. It occurs in two areas, and covers about a third of all unenclosed moorland. The largest area forms the centre of the northern moor, covering an area from about Yes Tor and High Willhays in the north, to Black Dunghill and Beardown Tors in the south, and about Manga Hill in the east to Great Links Tor in the west. Ornithologically it is the high plateau areas around Cranmere Pool, Cut Hill and Cowsic Head that are the most important, because it is here that the Golden Plover and Dunlin breed. The southern, smaller area of blanket bog extends over the moor from about Ter Hill to Stall Moor, and White Barrow to Lee Moor. Unlike the northern section it does not have regular breeding Dunlin and Golden Plover. Blanket bog vegetation is dominated by various grass species. In a recent survey this habitat was found to hold the highest breeding density of Skylarks, with additionally a high density of breeding Meadow Pipits. Where heather exists it is also important for Red Grouse (Geary 2001).

Upland Heathland

This habitat, which covers much of the open moor that is not covered by blanket bog, is dominated by heather species, with Western Gorse and Bilberry (Whortleberry) in places. Purple Moor Grass and Bristle Bent are also typical. It covers about 115 square kilometres of the moor. When overgrazed, upland heath degenerates into rough grass moorland. Much of Holne Moor, Haytor Down, Hameldown and Cosdon Beacon are examples of this habitat. Skylark and Meadow Pipit are typical breeding birds, along with Stonechat and Whinchat. Species of special interest are Red Grouse and Ring Ouzel. It is possible that a future coloniser could be the Dartford Warbler which has already been noted in some areas.

Lowland Heathland

There are only about three square kilometres of lowland heathland on Dartmoor. The habitat is not always easy to differentiate from upland heathland. The main areas are Roborough Down, Crownhill Down and Trendlebere Down. The main plants are heather and heath species with gorse, especially Western Gorse. The bird species have a slightly different emphasis, with Meadow Pipits still being found, but the number of Skylarks being much reduced. Stonechats can be plentiful and Linnets also, where gorse is plentiful. Yellowhammers can be in good numbers around the edges of certain areas where hedgerows occur, and Tree Pipits found where there are scattered small trees and bracken (Geary 2001). This is also the prime habitat for the increasing number of breeding Dartford Warblers.

Grass Moor and Bracken

Dartmoor has about 5.3 square kilometres of grass moor, with an additional 4.9 square kilometres of bracken. The grass moor is dominated by various moorland grass species, with grassland plants such as Heath Milkwort, Tormentil, and Heath Bedstraw. It is a product of heavily grazed heathland, and noteworthy areas occur in the south around Ugborough Beacon and Harford Moor, in the west around Cox and Staple Tors, and around the central basin at Royal Hill, Smith Hill and Littaford Tors. The Meadow Pipit reaches its highest Dartmoor breeding densities in this habitat, when the grazing is not heavy. With the Wheatear the opposite is the case, and high numbers can be reached in areas of heavily grazed grassland, with clitter and stone walls nearby (Geary 2001). Skylarks are also one of the main species in moderately grazed moor. Bracken is a highly invasive species that spreads onto grassland whenever it has the chance. Large areas are found around the edge of the moor, and unfortunately in recent years many new areas have been invaded. The Whinchat is one of relatively few species that benefit from bracken in the breeding season. Tree Pipits appear to benefit from its presence as long as there are scattered scrub and trees also present, and Yellowhammers, Linnets and Stonechats can also be found (Geary 2001).

Valley Mire

About 1.0 square kilometres of Dartmoor comprises of water-logged deep peat areas in valley bottoms, known as valley mire. This habitat, with acidic wetland plant communities, is unparalleled in upland Britain and makes Dartmoor internationally important. Examples of valley mire can be found in stretches of all moorland rivers, with particularly good areas at Taw Marsh, Raybarrow Pool, Foxtor Mires, and Halshanger and Bagtor Mires. Blackslade Mire and Whitchurch Down are designated Sites of Special Scientific Interest, purely because of their outstanding mire habitat. Plants include sedges and bog mosses, Bog Pimpernel, Bog Bean, Bog Asphodel, Pale Butterwort, Round Leaved Sundew, and Cotton Grass. It is the primary habitat for breeding Snipe, and is also important for the few remaining pairs of breeding Curlew. Lapwing also take advantage of these mires, particularly when there is short sward grassland for feeding close by. It is also a very important habitat for Dartmoor's Reed Bunting population, and Meadow Pipits can be found in good numbers.

Rhôs Pasture

This habitat, at times linked to valley mire, forms in valley systems away from the open moor. It occurs in small areas in many places around the moor but particularly around the north-east and east periphery. It is a rare and specialised habitat comprised of enclosed pastures with Purple Moor Grass and rushes, together with wet woodland, other grassland and at times oak wood. Although localised it forms the main area on the moor for Sedge and Grasshopper Warblers, is important for other warblers, and the willow carr areas are inhabited by Marsh Tits, and the few Willow Tits that breed around the moor. Spotted Flycatchers use the mature trees around the woodland edge, and Reed Buntings breed in the wetter, ranker parts. Wintering Snipe occur some years in numbers, as well as the small number that breed, and wintering Woodcock occur. The pastures provide excellent hunting territories for Barn Owls, and in winter the occasional Short-eared Owl or Hen Harrier may hunt.

Upland Oakwood

These woods occur above 250 metres and are dominated by Oak, although stands of Alder and Ash can be present. The habitat is of international importance and about 50 square kilometres occurs on Dartmoor. Birch and Beech can also be found, along with Rowan, Hazel and Holly in the understorey. Most of the upland oakwood in the area is classed as ancient, that is thought to have been in existence before 1600. Good examples of this woodland are Holne Chase and the Dart Valley woods, Yarner Wood and the adjacent Bovey Valley woods, Dunsford Wood and other woodland in the Teign valley. On a small scale the relict woodlands of Wistmans Wood, Black Tor Copse and Piles Copse, high up their respective river valleys, are also very important especially for their mosses, liverworts and lichens. Characteristic birds of these woodlands are the Redstart and Wood Warbler, together with the Pied Flycatcher which has increased in numbers through the use of nest boxes. Tree Pipits can be found on the edges of woodland, and Great Spotted Woodpeckers and Buzzards also inhabit the trees.

Wet Woodland

Although not a major habitat, about four square kilometres are to be found on Dartmoor, mainly in water-logged ground in valley systems, sometimes in association with oakwoods or Rhôs pasture. The normal trees are Willow and Alder. Breeding Willow Tits and Redpolls can be found in these areas, and roving flocks of Redpolls and Siskins feed in the Alders in winter.

Conifer Plantation

These plantations cover about 47 square kilometres of Dartmoor's uplands. The first were planted early in the last century. Much of the early timber has now been clear-felled, and in some cases replanted. Hence the mosaic of trees in various stages of growth, together with clear-fell areas, is a much more interesting prospect than the dense, dark, conifer blocks of 20 or 30 years ago. Birdlife is now much more diverse. There are several small areas around the moor, but the larger plantations at Bellever, Soussons, Fernworthy, Burrator and Beardown are known to most people. Although heartily disliked by many, or perhaps most, Dartmoor lovers, there is no doubt that over the years these plantations have been of great assistance to birdlife, and have provided Dartmoor with many breeding species that would otherwise have not occurred. In the early to mid-twentieth century, breeding Montagu's Harrier and Black Grouse were given a helping hand by the provision of young plantations. Merlins were assisted with nesting sites at the same, or possibly slightly later, stage. Today, although those species have gone, Hobbies and Goshawks are reliant to a large degree on the remaining mature stands. Crossbills and Siskins feed on the cones, especially Spruce and Larch, in the winter, and, in the case of the Crossbill, stay some years to breed. Siskins breed regularly now in all plantation areas. Young replanted areas attract Whinchats and Grasshopper Warblers when the trees are small, with a good growth of herbage around them, and when trees are a little bigger Redpolls nest. Dense populations of Coal Tits and Goldcrests breed in most plantations. Recently, the areas of clear-fell have been colonised by Nightjars in the summer. Numbers now breeding are a considerable proportion of Devon's total population. So although a plantation represents a blot on the landscape to the purist, it is also a habitat that can offer much to the observer of Dartmoor's birds.

Agricultural Land, Fields, and the Hedgerows containing them
Considerable agricultural land comes within the Dartmoor National Park, and there are about 4000 kilometres of hedgerows. Much of this land lies on the outskirts of the moor proper, especially on its eastern borders, but there is considerable incursion in the valleys of the River Webburn, around the prison fields at Princetown, smaller areas in the valley of the West Dart around Sherberton, Prince Hall, and Dunnabridge, and around the ancient tenement sites of Babeny and Sherril. Hay meadows and other dry grassland sites with particularly rich plant communities occur in certain areas. The birdlife of these agricultural habitats do not contain many special species, but two are of note. Dartmoor's small population of Woodlarks are located mainly in agricultural areas of the Teign valley. This is part of the only Woodlark population in Britain that breeds in this habitat. The Cirl Bunting, now expanding its British population in the South Hams, has reached the Teign valley also, where a few pairs breed, and can also be found occasionally in other moor edge farmland.

Tors, Quarries ,Spoil Heaps, and other Industrial Waste sites
Granite tors are a quintessential part of the Dartmoor landscape. The huge rocks crowning so many hills are perhaps of rather less value as breeding habitat for bird species now than they were in the past before the advent of tourism. Ravens, once common on the outcrops, now breed on only a few tors, usually as well away from human disturbance as is possible. However, the tors and the clitter that surround many of them are still important nesting sites for Wheatear and some Ring Ouzels. Spoil heaps, from ancient open cast mining, also offer habitat for the latter two species. Spoil heaps from another Dartmoor industry, china clay, and the adjacent pits, are also useful for Wheatears, and at times a land slip will reveal a sandy bank on

a spoil heap or nearby pit that will allow a Sand Martin colony to become established. Quarries, usually around the moor edge, remain important sites for Ravens and certain raptor species.

Buildings

Even in a rural habitat like Dartmoor human habitation can be very important for some species. House Martins, Swallows and Swifts rely on buildings almost totally for nest sites, whether it is village or town houses, or farms and their outbuildings. Jackdaws are drawn to chimney sites, and are a nuisance to many householders in Dartmoor towns. House Sparrows, of course, have a long history of sharing habitation with man, and the Barn Owl is still very much reliant on man for nesting sites, in its territories around the moorland fringe.

Rivers and Streams

Dartmoor is a land of rivers. Because they are acidic and fast flowing, they are unable to support many water plants, and the avifauna tends to be rather specialised and more restricted than on the slow flowing rivers of the lowlands. Dippers and Grey Wagtails are associated with just about any sizeable river on the moor, as well as many of the smaller streams. They are mostly found on the middle courses of rivers, when the streams have left their high moorland sources, and gain size and momentum as they descend to the moor edge and "in country". On the lower levels of some rivers, notably the Dart, Kingfishers can be found on slower flowing stretches. Grey Herons can be found fishing on most rivers, sometimes almost up to the source. Two duck species have quite recently spread to Dartmoor waterways to breed. They are the Goosander and the Mandarin. Both can be found in small numbers on certain rivers, notably again the Dart, where holes in banks and nearby trees provide important nest sites.

Reservoirs and Ponds

Dartmoor has no natural lakes, and the nearest that we have to large open stretches of water are the reservoirs built from the late nineteenth to the mid-twentieth centuries to provide drinking water for Devon's towns. The reservoirs of Burrator, Fernworthy, Meldon, Venford, Hennock and Avon are known to most Dartmoor visitors, and prove great attractions. As they are acidic and mainly deep and steep sided, they are of somewhat limited attraction to bird life. They do not normally possess much in the way of muddy margins, so are not favoured by many waders on passage. Although Mallard breed regularly on most, and Teal rarely on one, ducks generally tend to be few. But the open water has proved very useful for the numbers of Goosanders that roost communally on reservoirs during the winter months, and they are important in the late summer and autumn for roosting Canada Geese. Great Crested Grebes have gradually begun colonising the reservoirs in recent years. Ponds, or more correctly settling ponds in the china clay district of Lee Moor have provided important habitat in the last thirty or so years to Dartmoor's only breeding Common Sandpipers. This is also the site of the county's only records of breeding Little Ringed Plover. Other smaller ponds on the moor, such as Knattabarrow Pool, the pools on Crownhill Down, and Cramber Tor Pool, are important for insects, especially dragonflies, and plants but tend to be too small for many bird species, although they can be interesting during periods of migration, when scarce species can occasionally turn up.

A DARTMOOR YEAR

January

A New Year, and birdlife on Dartmoor is about to endure the two coldest months of the year. A great deal depends on the severity of the weather. Hen Harriers drifting in to roost near Warren House Inn each evening regularly reach their highest numbers this month; but if Arctic weather hits the region, birds will disperse to lower areas. In severe conditions Golden Plover flocks will desert the moor very quickly. If their favoured short grass sward feeding areas are frozen, they must move at once or starve. With autumn berries now almost depleted, Redwings and Fieldfares take to cultivated fields for sustenance. Frozen conditions, and they too must leave the moor. So much then depends on the weather this month. In mild conditions many species can be present, in severe conditions the uplands can be all but deserted. One species that is usually seen in increased numbers on certain reservoirs at this time is the Goosander. Good numbers come in to roost in the late afternoons, particularly at Venford and Burrator. Meadow Pipits and Skylarks are almost completely absent from the breeding areas. But to show that Spring is not too far away Ravens start to exhibit breeding activity, and in the plantations, if any have stayed since the previous autumn, Crossbills may be showing signs of breeding, with red males singing in the Spruce stands.

February

Another cold month on the moor, and yet if a few mild days occur, perhaps with low cloud, Skylarks can be heard overhead as they fly back from the lowlands. The first song, indicating the claiming of territories, can be heard some years quite early in the month, if a mild period persists. Ravens are now nesting in earnest, and if you are lucky you may witness early Goshawk display over the plantations. Dippers sing as territories are reclaimed on rivers, or quite possibly just become more vocal on stretches where they have been present all winter. In Hembury Woods, Lesser Spotted Woodpeckers begin their spring courtship with drumming and display.

March

Early in March further signs of the changing seasons can be seen. The first Meadow Pipits are back on territories early in the month, although the total moorland breeding population takes several more weeks to return, and migrant flocks can still appear in early April. The first Wheatears are always eagerly awaited, and most years early males have returned by the first or second weeks of the month. Early Sand Martins arrive back from the south about the same time. The bubbling calls of Curlew can be heard from about mid-month in the few marshes still occupied, and Lapwing are back wheeling and calling over their territories. Winter visitors, in the meantime, are becoming fewer. The Hen Harrier roost near Warren House Inn now has few visitors, as many birds have moved north, and most Redwings have gone, although Fieldfares tend to stay a little longer. Stonechats and Reed Buntings move back onto breeding territories by the latter part of the month, and by the end of the month Grey and Pied Wagtails have moved back onto streams from their winter quarters off the moor. By the months end, Chiffchaffs can also be heard singing in all likely habitat, and the first Ring Ouzel may have appeared in the mine gullies at Vitifer.

April

A month of contrasts. Early in the month it is not at all exceptional for the last bite of winter to be felt. At times snow remains on the hills for a day or two. Later in the month the first real warmth of spring arrives. Wheatears, still arriving during the month, are joined by Redstarts and Tree Pipits from mid-month, and Whinchats a little later. Willow Warblers, our commonest warbler, arrive back into scrub and plantation sites from early in the month, with their close relative the Wood Warbler from mid-month, about the same time as the first Pied Flycatchers. Dartford Warblers sing their hurried, scratchy song from gorse on heaths and downs, and Swallows return to Dartmoor farms. The Cuckoo, now much more regular on the moor than in lowland regions, starts to arrive from the third week of April. The last of the wintering Golden Plover leave to go north, and are replaced for a short time by migrant flocks that have wintered further south, passing through on their way north. It is at this time, if you are very fortunate, that you may come across a Dotterel on some high grass moorland, or even a small "trip".

May

Most summer migrants have arrived by mid-month, and breeding is now in earnest. On the high northern moor Golden Plover and Dunlin have taken up territories, and are vocal early in the mornings, when their songs echo during flights over the lonely blanket bog. Male Pied Flycatchers are now back in numbers, and at Yarner, Okehampton and elsewhere, are singing almost continuously high in the oaks. Hobbies have arrived, and although they can be very secretive near nest sites, excellent views can be obtained on warm and sunny afternoons and evenings, as they hunt for moths over heather moorland. Goosanders that have bred on the rivers usually have young by now, and a quiet approach can at times give good views of a female with up to a dozen ducklings. A quiet walk in almost any of the plantations at dusk in late May onwards may detect the strange prolonged "churring" of Nightjars. As most of the plantation areas have been partially clear-felled in recent years, large areas are available to this species, and numbers on the moor have increased greatly. Late May and early June is a very good time, if you have luck, to find over-shooting spring migrants that should by rights have remained further to the south. Red Kites, Red-backed Shrikes, and Montagu's Harriers have occurred recently at this time, as have extreme rarities such as Rock Thrush and Spectacled Warbler.

June

Many resident species now have young on the wing, and some of the early migrants, especially Wheatear, have fledged broods. The maximum numbers of Nightjars have by now arrived on territories, and if Great Crested Grebes have bred at Fernworthy, young are usually seen late in the month. For the patient observer, proof of breeding for Dunlin and Golden Plover may be obtained if young are seen or recently discarded egg shells found, but they are elusive and difficult, and evidence of probable breeding is the most that can usually be confirmed. Throughout the month, Rooks with their families visit grass moor to feed on Crane Fly larvae. A count in the Cox Tor or Walkham Valley areas can sometimes reveal that many hundreds are present. Common Sandpipers, if they have bred on the china clay pits, disperse onto nearby rivers and reservoirs in late June, and are joined by dispersing breeding birds from further afield. Also at the end of the month there is the chance of seeing Crossbills. If birds have irrupted from their Scandinavian breeding areas, this is the usual time when the first waves of migrants reach our plantations. If there has been a good cone crop, birds will stay around, and then the trees at Fernworthy or Bellever are filled with the "chipping" calls of these birds.

July

Young fledglings are everywhere this month. Bird song is almost totally absent, and all habitats are filled with family parties, with the young either still being fed by adults or just having gained their independence, beginning to feed themselves. There is little movement away from the moor, although it is likely that any breeding Golden Plover and Dunlin will quietly depart during the latter part of the month. As the month progresses a few waders, notably Green and Common Sandpipers and perhaps Greenshank, appear on water courses and reservoirs. These are non or failed breeders that have departed early from breeding grounds to the north, or Scandinavia. In invasion years Crossbill numbers may continue to build up, or the woods may go quiet again as earlier parties move away. Much depends on the state of the Dartmoor cone crop.

August

This is the month of departing birds. Most of the summer migrants go south during this period. Departure is quiet and little recorded, but any observer will note that during August, the number of birds recorded will decrease. Wheatears become fewer, Redstarts disappear, many warbler species will go from breeding sites and be found in atypical habitat as they disperse and build up their fat reserves prior to migration. Most Siskins will have left their breeding plantations by the end of the month. Wader passage continues with the same species that occurred in July, sometimes augmented by the occasional Dunlin and Ringed Plover. Even common residents tend to move off the moor. Skylarks become scarce by mid-month, and most Meadow Pipits have also moved away. Late in the month young Hobbies can become quite vocal as they sit around in the spruce waiting for parents to feed them.

September

Meadow Pipits, scarce in August, appear again this month, sometimes in good sized flocks, feeding amongst long grass, heather and the brown autumn bracken. These are largely migrants from further north, coming off passage to feed. Merlins, taking advantage of the food source, often move south with them. Occasionally a rarity, such as a Richard's or Tawny Pipit, will be noted with the migrating pipit flocks. After their late summer moult, numbers of Canada Geese begin to increase on the reservoirs, especially Fernworthy. Ring Ouzels appear around the moor edge, sometimes in small flocks, feeding on the Rowan berries. Migrant Wheatears are still to be found feeding in areas of short grass, but as the last warm days go by, a chill is noticeable morning and evening and summer gives way to autumn.

October

The first of the wintering Hen Harriers return early in the month, and it is possible that a bird hunting over the heather near Warren House Inn or Birch Tor prior to roosting, may have a Merlin in attendance, waiting to dash in and take any bird that the more ponderous Harrier flushes. Golden Plover flocks increase greatly at this time and numbers in the Cox and Staple Tors area can soon reach several hundred. Also by mid-month the first of the winter thrushes begin to arrive from Scandinavia. Redwings are almost invariably the first to appear, although their numbers can vary greatly from year to year. Fieldfares tend to arrive in late October or early November and usually in greater numbers than the earlier Redwing. Together they strip the hawthorns of berries before moving to another location. Interestingly at this time also a few late Ring Ouzels can also be found with the thrushes. These are Scandinavian bred birds moving with their more numerous relatives. Jays are easier to see at this time than any other time of the year, as they fly to and fro collecting acorns to secrete for later consumption.

November

As the damp and raw days of November take hold, birdlife prepares for winter. The wintering continental Starling flocks have arrived and in the late afternoons numbers build up as they prepare to roost. Larger numbers over-fly the moor on their way to the huge roost just west of our area at Thorndon Cross. There could well be a sprinkling of Black Redstarts in likely urban areas, or around tors early in the month, and the first wintering birds arrive in the gardens at Buckfastleigh. Although not a time when too much ornithological excitement can be expected, a passing Short Eared Owl is possible, and perhaps a Great Grey Shrike will be returning to winter in the Vitifer area.

December

Days shorten and get colder. Most birds are now off the high northern moor, only the Red Grouse stay to tolerate whatever the winter weather brings. Flocks of Golden

Plover will stay in their traditional short turf areas around the moor edge as long as the weather is not severe. Occasionally a Peregrine will sweep low in an attempt to take a bird from the startled throng. A walk through any boggy ground may flush several Snipe, and at times a Jack Snipe will fly up almost at the walker's feet. Large parties of Ravens gather around the rocks of Cox Tor in the afternoons for pre-roost gatherings, flying in pairs high above the tor, or sitting together amongst the granite boulders. If the beech mast crop has been good, Bramblings will gather with Chaffinches to feed at likely sites at Fernworthy and Postbridge. Whilst driving over the moor on crisp, cold evenings, perhaps after watching the Hen Harriers coming into roost, it is not unusual to see several Woodcock flighting out of the plantations at Bellever to feed by night in the small fields near Postbridge.

BIBLIOGRAPHY AND REFERENCES

General Published Works
AVERY Mark and LESLIE Roderick 1990. *Birds and Forestry.* T. & A.D. Poyser, London.
BALLANCE David K. and GIBBS Brian D. 2003 *The Birds of Exmoor and the Quantocks.* Isabelline Books, Falmouth.
CAMPBELL Bruce and FERGUSON-LEES James 1972. *A Field Guide to Birds Nests.* Constable, London.
CRAMP Stanley, et al 1977 to 1994. *Handbook of the Birds of Europe, the Middle East and North Africa, 9 Volumes.* Oxford University Press, Oxford.
FULLER R.J. 1982. *Bird Habitats in Britain.* T. & A.D. Poyser, Calton.
GIBBONS D.W. et al 1993. *The New Atlas of Breeding Birds in Britain and Ireland 1988-1991.* T. & A.D. Poyser, London.
HOLLOWAY Simon 1996. *The Historical Atlas of Breeding Birds in Britain and Ireland.* T. & A.D. Poyser, London.
LACK Peter 1986. *The Atlas of Wintering Birds in Britain and Ireland.* T. & A.D. Poyser, Calton.
MEAD Chris 2000. *The State of the Nations Birds.* Whittet Books, Stowmarket.
NETHERSOLE-THOMPSON Desmond and Maimie 1986. *Waders, their Breeding, Haunts and Watchers.* T. & A.D. Poyser, Calton
PALMER Philip 2000. *First for Britain and Ireland 1600-1999.* Arlequin Press, Chelmsford.
RATCLIFFE Derek 1990. *Bird Life of Mountain and Upland.* Cambridge University Press, Cambridge.
RATCLIFFE Derek 1997. *The Raven.* T & A.D. Poyser, London.
SHARROCK J.T.R. 1976. *The Atlas of Breeding Birds in Britain and Ireland.* British Trust for Ornithology, Tring.
TYLER Stephanie and ORMEROD Stephen 1994. *The Dippers.* T. & A.D. Poyser, London.
WATSON Donald 1977. *The Hen Harrier.* T. & A.D. Poyser, Berkhamsted.
WERNHAM Chris et al 2002. *The Migration Atlas: Movements of the Birds of Britain and Ireland.* T.& A.D. Poyser, London

Published Works relating, at least in part, to Dartmoor
BELLAMY J.C. 1839. *The Natural History of South Devon.* Plymouth and London.
BRAY Mrs. A. E. 1836. *A Description of the Part of Devonshire bordering on the Tamar and Tavy.* John Murray,London.
BROWN Mike 1995. *The Gazetteer of Dartmoor Names.* Forest Publishing, Newton Abbot.
CHOWN D. et al 1992. *Dartmoor Environmental Baseline Breeding Bird Survey.* RSPB.
CROSSING William 1912. *Guide to Dartmoor.* Western Morning News, Plymouth.

D'URBAN W.S.H. and MATHEW the Rev. M.A. 1892. *The Birds of Devon.* R.H. Porter, London.

D'URBAN W.S.H. 1906. *The Victoria County History of Devonshire.* London. The section on Birds.

ENGLISH NATURE/DNPA 2001. *The Nature of Dartmoor : A Biodiversity Profile.*

ENGLISH NATURE/DNPA 2001. *Action for Wildlife : The Dartmoor Biodiversity Action Plan.*

FIELD STUDIES COUNCIL 1979. *Vegetation Map of Dartmoor.*

GEARY Simon et al 2000. *Dartmoor Moorland Breeding Bird Survey.* DNPA, Bovey Tracey.

GORDON Douglas 1931. *Dartmoor in all its Moods.* John Murray, London.

HAMILTON-LEGGETT Peter 1992. *The Dartmoor Bibliography.* Devon Books, Tiverton.

HARVEY L.A. and ST.LEGER-GORDON D. 1953. *Dartmoor.* New Naturalist, Collins, London.

HEMERY Eric 1983. *High Dartmoor : Land and People.* Robert Hale, London.

JENKS David G. 2004. *A History of Devonshire Ornithology.* Isabelline Books, Falmouth.

MOORE Robert 1969. *The Birds of Devon.* David and Charles, Newton Abbot.

MUDGE G.P. et al 1979. *An Ecological Study of Breeding Bird Populations and Vegetation on Open Moorland Areas of Dartmoor.* RSPB/DNPA.

NORMAN David and TUCKER Vic 1984. *Where to Watch Birds in Devon and Cornwall.* Helm, London.

PIDSLEY W.E.H. 1891. *The Birds of Devonshire.* Gibbins, London and Exeter.

ROBINS M. and JUTSUM R. 1986. *Dartmoor Environmental Base Line 1986 Breeding Bird Survey.* RSPB.

ROSIER A. et al 1995. *A Checklist of the Birds of Devon.* DBWPS.

ROWE Samuel 1848. *A Perambulation of the Antient and Royal Forest of Dartmoor.* Plymouth and London.

SITTERS H.P. 1974. *Atlas of Breeding Birds in Devon.* DBWPS.

SITTERS H.P. 1988. *Tetrad Atlas of the Breeding Birds of Devon.* DBWPS.

SWINFEN Roger 1979. *Checklist of the Birds of Dartmoor.* DBWPS.

WALMESLEY WHITE W. *The Ornithologcal Journal.* 2004. Isabelline Books, Falmouth.

Journals and articles in Journals relating to birds on Dartmoor

CHESHER Jeremy 1996. *Pecking Order.* Dartmoor Bird Report 1996.

DARTMOOR STUDY GROUP. *Dartmoor Bird Report.* 1996-2002.

DEVON BIRD WATCHING AND PRESERVATION SOCIETY. *Annual Reports.* 1929 - 2002.

ELLICOTT P.W. and MADGE S.G. 1960. *Redstart Distribution Enquiry.* 33rd Report of Devon Bird Watching and Preservation Society

ELLICOTT P.W. *Redstart Distribution Enquiry.* 41st and 42nd Reports of Devon Bird Watching and Preservation Society.

GEARY Simon 2000. *Dartmoor Moorland Breeding Bird Survey 2000.* 73rd Report of Devon Bird Watching and Preservation Society.

GOODFELLOW P.F. *Owl Enquiry.* 32nd to 37th Reports of the Devon Bird Watching and Preservation Society 1959-1964.

HIBBERT Richard 2000. *Huccaby Bird Report.* Dartmoor Bird Report 2000.

HIBBERT Richard 2001. *The Second Huccaby Bird Report.* Dartmoor Bird Report 2001.

HIBBERT Richard 2002. *Breeding Curlew and Lapwing, Eastern Dartmoor 2002.* Dartmoor Bird Report 2002.

HURRELL H.G. 1947. *The Birds and Mammals of Dartmoor.* Transactions of the Plymouth Institution, Vol. 20 1945-47.

HURRELL H.G. 1956. *A Raven roost in Devon.* British Birds, Vol. XLIX No. 1 1956.

HURRELL H.G. 1980. *A Large Gathering of Cuckoos.* British Birds Vol. 73 1980.

JONES Bob 1999. *Golden Plovers & Dunlins breeding on Dartmoor.* Dartmoor Bird Report 1999.

LANCUM F. Howard 1934. *Merlins Breeding on Dartmoor.* British Birds Vol. XXV111 No. 5 1934.

ST. LEGER John 1996. *Summary of Golden Plover and Dunlin breeding records, Dartmoor 1996.* Dartmoor Bird Report 1996.

SAMPSON Mike and ST.LEGER John 1997. *Golden Plover and Dunlin Breeding on Dartmoor in 1997.* Dartmoor Bird Report 1997.

SMALDON Roger 1996a. *A Summary of the Results of the Roosting Goosander Survey in the winters of 1993/94 and 1994/95.* Dartmoor Bird Report 1996.

SMALDON Roger 1996b. *Merlins Hovering - Further Comment.* Dartmoor Bird Report 1996.

SMALDON Roger 1997a. *Observations at the Birch Tor Hen Harrier Roost.* Dartmoor Bird Report 1997.

SMALDON Roger 1997b. *In years gone by ...* Dartmoor Bird Report 1997.

SMALDON Roger 1997c. *Juvenile Peregrine defending territory in its first autumn.* Dartmoor Bird Report 1997.

SMALDON Roger 1998a. *Golden Plover and Dunlin breeding on Dartmoor in 1998.* Dartmoor Bird Report 1998.

SMALDON Roger 1998b. *The Birch Tor Hen Harrier Roost 1998.* Dartmoor Bird Report 1998.

SMALDON Roger 2002a. *Hen Harriers roosting on Dartmoor in 2002.* Dartmoor Bird Report 2002.

SMALDON Roger 2002b. *Nightjars on Dartmoor 1997-2002.* Dartmoor Bird Report 2002.

WALMESLEY WHITE W. 1934. *Merlins breeding on Dartmoor.* British Birds Vol. XXV111 1934.

The following articles are all to be found in DEVON BIRDS the Journal of the Devon Bird Watching and Preservation Society:

ARCHER-LOCK A. and SLADE L.W. 1968. *The Birds of High Dartmoor.* Vol. XXI No.4 1968.

BEER T.V. 1949. *A Party of Ring Ouzels.* Vol. II No.2 1949.

BUBEAR David 1973. *Nest Boxes in Meldon Wood.* Vol. XXVI No. 3 1973.

DARE P.J. 1957. *The Post-Myxomatosis Diet of the Buzzard.* Vol. X No.1 1957.

DARE P.J. 1958. *Birds of the Postbridge Area.* Vol. XI No.2 1958.

DARE P.J. 1959. *The Birds of the Postbridge Area 1955-58. Additional Notes.* Vol. XII No.1 1959.

DARE P.J. and HAMILTON L . I. 1964. *Birdlife at Postbridge during winter 1962/63.* Vol. XVII No.1 1964.

DARE P.J. and HAMILTON L.I. 1968. *Birds of the Postbridge area, Dartmoor 1956-1967.* Vol. XXI Nos. 2 and 4 1968.

DARE P.J. 1969. *Hovering by the Common Buzzard.* Vol. XXII No. 4 1969.

DARE P.J. 1996. *Birds of the Postbridge area 1955-1994.* Vol. 49 No. 2 1996.

DARE P.J. 1998. *A Buzzard Population on Dartmoor 1955-1993.* Vol. 51 No.1 1998.

DARE P.J. 1999. *Large Movements and Ground Assemblies of Buzzards in Devon.* Vol. 52 No.2 1999.

EASTON Gary and Anna 1996. *Swallows nesting in old mineshaft on Dartmoor.* Vol.49 No.2 1996.

GOODFELLOW P.F. 1958. *Close-up of Grasshopper Warblers.* Vol. XI No. 1 1958.

GOODFELLOW P.F. 1966. *The Owl Enquiry, Part 1, The Barn Owl.* Vol. XIX No.2 1966.

GOODFELLOW P.F. 1986. *The Wood Warbler Survey 1984-85.* Vol. 39 No. 4 1986.

GOODFELLOW P.F. 1987. *Wood Warblers - 60 years ago.* Vol. 40 No. 2 1987.

HUBBLE Rob 1993. *Aggression between juvenile Dippers.* Vol. 46 No. 1 1993.

HURRELL H.G. 1950. *The Kite in Devon.* Vol.3 Nos. 1 and 2 1950.

HURRELL H.G. 1960. *The Kite in Devon.* Vol. 13 No.3 1960.

HURRELL H.G. 1974. *The White Buzzard of Buckfastleigh.* Vol. XXVII No.1 1974.

HURRELL Kenneth 1963. *Dartmoor Naturalists take to Skis.* Vol. XVI No. 3 1963.

HURRELL Leonard 1951. *Golden Plover and Dunlin on Dartmoor.* Vol. 4 No. 1 1951.

JENKS Dave 1997. *The 1996 Sample Census of Rookeries.* Vol. 50 No. 1 1997.

JENKS Dave 2000. *Survey of Devon Rookeries 1996.* Vol.53 No.2 2000.

JOHN Tony 1990. *Icelandic Merlin in Devon.* Vol. 43 No. 3 1990.

JONES Bob 1989. *An Unusual Winter Visitor.* Vol. 42 Nos. 2/3 1989.

JONES Bob 1996. *A Study of Ring Ouzels Breeding on Dartmoor.* Vol. 49 No.2 1996.

JONES Bob 2001. *Golden Plovers and Dunlins Breeding on Dartmoor.* Vol. 54 No.1 2001.

JONES Bob 2004. *Movement of Golden Plovers Breeding on Dartmoor.* Vol. 57 No. 1 2004.

JONES John and Janet 1980. *Redstarts at Burrator.* Vol. 33 No.2 1980.

ST.LEGER John 1990. *Search for Breeding Dunlin and Golden Plover on N. Dartmoor.* Vol. 43 No. 1 1990.

LEONARD E. et al 1977. *Birds in Wistmans Wood, Dartmoor.* Vol. XXX No. 4 1977.

McEWEN Iris 1981. *Black and White Warbler at Tavistock - A New Bird for Devon.* Vol. 34 No. 2 1981.

McMAHON Sarah 1994. *The Birds of Portworthy.* Vol. 47 No. 2 1994.

MUDGE G.P. et al 1981. *Breeding Bird Populations of the Open Moorland of Dartmoor in 1979.* Vol. 34 No. 2 1981.

NELDER J.A. 1949. *Golden Plover and Dunlin on Dartmoor.* Vol. 11 No. 3 1949.

NILES John 1971. *High Altitude Dunnocks.* Vol. XXXIV No. 2 1971.

PAGE Phil 1992. *Yarner Wood NNR and its Birds - Forty Year On.* Vol. 45 No. 2 1992.

ROBINSON W. 1949. *Montagu's Harrier.* Vol.11 No. 3 1949.

SALMON Paul 1988. *A National Park Ranger's Hobby.* Vol 41 No. 1 1988.

SITTERS H.P. 1975. *1974 Tetrad Survey of Stonechat and Reed Bunting.* Vol. XXXVIII No. 1 1975.

1974 Tetrad Survey of Yellowhammer, Cirl Bunting, and Woodlark. Vol. XXXVIII No. 2 1975.

1974 Tetrad Survey of Whinchat, Wheatear and Redstart. Vol. XXXVIII No.4 1975.

SMALDON Roger 1982. *Birds of Burrator.* Vol. 35 No. 2 1982.

Diurnal Haunts of Burrator's Goosanders. Vol. 35 No. 4 1982

SMALDON Roger 1993. *Summer Flocking of Juvenile Starlings on Dartmoor.* Vol. 46 No. 1 1993.

Goosanders Roosting on Dartmoor Reservoirs in the Winter of 1992/93. Vol. 46 No. 2 1993.

SMALDON Roger 1994. *The Breeding Birds of Dartmoor's Relict Oakwoods.* Vol. 47 No. 1 1994.

SMITH F.R. 1956. *Montagu's Harrier in Devon.* Vol. IX No. 3 1956.

VAUGHAN Gordon 1979. *The Pied Flycatchers of Okehampton.* Vol. XXXII No.2 1979.

VAUGHAN Gordon 1981. *The Pied Flycatchers of Okehampton - a Progress Report.* Vol. 34 No. 2 1981.

VAUGHAN Gordon 1989. *Dormousitis - how virulent a disease is it becoming?* Vol. 42 No. 1 1989.

VAUGHAN Gordon 1994. *Pied Flycatchers in Okehampton - the First Twenty Years.* Vol. 47 No.1 1994.

VAUGHAN Gordon 2001. *Dormousitis - the Sequel.* Vol. 54 No. 2 2001.

WARE E.H. 1948. *The Pied Pipers of Dartmoor.* No. 9 1948.

WYNNE-EDWARDS V.C. 1979. *How Red Grouse Keep their Numbers in Balance.* Vol. XXXII No. 1 1979.

SYSTEMATIC LIST OF SPECIES

Classification
The order used in most County Reports and publications is followed here. This is the British Ornithologists' Union's *The British List* (Tring 2000). However the "new" English names in this List have not been adopted.

Dates
The cut-off point for the bulk of records in this publication was 31 December 2002. However, when particularly relevant, some records from 2003 have been included, and on the few occasions important 2004 records have been used.

References
As mentioned earlier, most records included have come from the annual reports of the Devon Bird Watching and Preservation Society, and latterly the Dartmoor Study Group. References to particular reports are not generally given, although records for species that have occurred on only ten or so occasions are given in detail. References to earlier sightings, plus all references in journals and survey reports, are given in the text, and fully listed in the References and Bibliography Section. Abbreviations have been kept to a minimum, but the following will be found:

BTO = British Trust for Ornithology
DNPA = Dartmoor National Park Authority
DBWPS = Devon Bird Watching and Preservation Society
RSPB = Royal Society for the Protection of Birds

Status Descriptions
The descriptions given apply only to Dartmoor, and may well be very different to Devon as a whole. Many species on Dartmoor have received only scant attention in the past, and with the exception of certain restricted areas and some professional surveys, still remain somewhat under-observed. With this in mind, for most species their status is given in simple terms e.g. passage migrant, resident breeder, etc., with no attempt at quantification. Where a species is of known rarity status the description is qualified. Likewise with known scarce species. Most of the descriptions of status follow the pattern used in most County Bird Reports. Where a species comes within more than one heading, its status will be described in as much detail as can be given without doubt. Most status descriptions will be self explanatory, but three require a little explanation. The term vagrant, either rare or very rare, refers to species that are rare on Dartmoor, but also rare nationally. Accidental visitor has been used to describe a species that may not be scarce or rare on a national or even county basis, but is completely exceptional on Dartmoor i.e. sea birds that are not known normally to over-fly land turning up after autumn or

winter gales. The term "altitudinal migrant" has been used to describe species that move off the high moor in winter, although perhaps go no further than the Devon lowlands. An example of an almost complete altitudinal migrant on Dartmoor is the Skylark, with the Stonechat being a partial one.

Place-names

With an area the size of Dartmoor place-names are many and varied. At times birds are reported from areas that are obscure and difficult to find on a map. Many of the more usual names will be found on Map 1 in this volume (see fold-out following p. 231), and almost all of them can be located on the Ordnance Survey 1:25 000 Dartmoor Explorer OL28 Map. However one publication that anyone interested in any aspect of Dartmoor should possess is *The Gazetteer of Dartmoor Names* by Mike Brown (Forest 1995) with later editions. This gives all Dartmoor place-names with six or eight figure map references and is completely indispensible.

RED-THROATED DIVER *Gavia stellata*
Rare winter visitor and passage migrant.
There are ten records of single birds.

1929:	Burrator, 23 March, one and presumably the same bird on 4 April
1950:	Burrator, 14 February to 15 March, one seen on several dates
1955:	Burrator, 8 January, one
1958:	Burrator, 7 April, one
1972:	Burrator, 13 February, one
1980:	Venford Reservoir, 17 February, one
1984:	Trenchford Reservoir, 17 March, one
1985:	Hennock Reservoir, 16 February, one
	Venford Reseroir, 20 February to 9 April, one
1986:	Burrator, 16 to 22 February, one badly oiled

Burrator is the most favoured location for this, as for all diver species records, As only one of the birds was seen in January, it is probable that all the remaining records refer to birds returning north to breeding grounds. The March and April records are obviously spring migrants, and the five February records could also refer to migrants, as the British breeding population is known to move back north in February, ahead of Scandinavian birds (Lack 1986).

The bird seen at Hennock Reservoir on 16 February 1985 was seen on the grass bank, which is usually a sign that the bird is sick or oiled. This may well have been the same individual that a few days later was found at Venford, and made a protracted stay until early April.

BLACK-THROATED DIVER *G. arctica*
Rare winter visitor and passage migrant.
There are seven records of single birds.

1982:	Burrator, 4 to 6 May, one in breeding plumage
1983:	Burrator, 27 February, one
1987:	Fernworthy Reservoir, 4 to 7 December, one
1993:	Burrator, 1 March to early June, one, moulting into breeding plumage
1996:	Burrator, 1 January to 18 March, one in first winter plumage
	Burrator, 14 July, one
	Burrator, 15 to 23 November, one

As all the records refer to individuals arriving in different months there is little pattern with their occurrences. Three could be birds diverted by bad weather from the wintering grounds on the coast and three spring migrants. The mid-July bird at Burrator is almost unexplainable. The long stay of the birds at Burrator in 1993 and 1996 was noteworthy.

GREAT NORTHERN DIVER *G. immer*
Rare winter visitor and passage migrant.
There are ten records of single birds.
> 1952: Burrator, 30 December to 10 January 1953, one
> 1961: Burrator, 15 December, one
> 1968: Fernworthy Reservoir, 14 December, one
> 1978: Burrator, 14 January to 14 February, one
> 1983: Fernworthy Reservoir, 6 March, one
> 1989: Burrator, 27 November to 15 December, one
> 1993: Venford Reservoir, 5 and 6 of June, one in breeding plumage
> 1996: Burrator, 13 January to 9 February, one in first winter plumage
> 1997: Burrator, 30 November, one
> 1999: Burrator, 8 to 14 January, one, and what was probably the same bird on
> 19 February

The majority of the records refer to birds seen during the winter months, no doubt blown off course by severe weather. The June record at Venford in 1993 is intriguing. Birds are found on the coast at times in June, so perhaps this was a late migrant taking an over-land short-cut back to the breeding grounds.

The 1952/53 bird at Burrator was interesting in as much as it was in full summer plumage despite the time of year. On the 10 January 1953 it was heard to give its wailing cries. The bird at Burrator in January 1996 was present at the same time as the Black-Throated Diver. This was almost certainly the only time on Dartmoor that the two species of diver have been present together.

LITTLE GREBE *Tachybaptus ruficollis*
Very scarce winter visitor and passage migrant. Has bred.
This is a bird of lowland Britain that does not do well in Devon. Its required habitat of slow flowing rivers and ponds with aquatic vegetation is in very short supply in the county, and almost non-existent on Dartmoor, with its acidic water and fast streams. At Horsham Pond, Mannaton, a pair was noted on 5 May 2000 collecting nesting material. Subsequently a bird was observed taking food to a probable nest site, and three juveniles were present on 11 July with the parents. This constituted the first known breeding on Dartmoor. The pair bred again in 2001, raising two young. The pair also bred at Buckfast in 2000 raising three young.

Occasionally birds are seen in spring on Dartmoor reservoirs. In 1994 two were heard calling at Burrator on 18 April, and the same year one was calling at Fernworthy on 9 May. Calling at this time of year in likely habitat often signifies a breeding attempt, but in both of the above cases nothing more was seen.

After the breeding season, birds disperse quickly to larger waters and estuaries, and this is when Dartmoor records increase. The earliest post-breeding individuals were two juveniles at Burrator on 25 July 1999, but Fernworthy Reservoir is the spot that attracts birds on a regular basis at this time of year. One or two are here most years from late July throughout the autumn, with the maximum being six on 7 September 2002. Birds are also noted occasionally at this time of year at Kennick Reservoir, and at china clay pools at Lee Moor and Portworthy. During the winter months one or two birds can be found occasionally at most reservoirs with Fernworthy and Kennick being the most used sites.

GREAT CRESTED GREBE *Podiceps cristatus*
Rare breeder, and very scarce passage migrant and winter visitor.
As with the previous species, largely a bird of the lowlands.
On Dartmoor it was unknown until the building of the reservoirs, and even then it did not put in an appearance until the mid-twentieth century.

The first record was of a bird on Burrator Reservoir on 16 February 1959. There were then no sightings for a further ten years, until another was seen at the same locality on 5 March 1969. Up until 1985, only two further birds appeared, both at Burrator.

The establishment of breeding birds in the county began with pairs at Slapton in 1973 (Sitters 1988), and there was a slow expansion onto other lowland sites from that time.

On Dartmoor, records become rather more frequent during the latter part of the 1980s, with winter records from Burrator, Kennick and Fernworthy Reservoirs, and a spring bird at Fernworthy on 10 May 1985. It was 1990 before the first signs of possible breeding occurred when a pair arrived at Fernworthy on 15 March. Nothing happened this year, however, and it was not until 1992 that a pair nested. This attempt failed however, due to a drastic change in water levels and the abandonment of the nest. The rapid variation of water levels is constant threat to the successful breeding of this species in upland areas (Gibbons et al 1993). Birds returned in 1993 and subsequent years, but it was not until 1996 that a pair successfully reared one young.

In the meantime, a pair took up residence at Kennick Reservoir in 1993, and claimed the first successful Dartmoor breeding record when they raised four young. They were successful again a Kennick in 1994 and 1995, raising four young each year, but had less luck in subsequent years, although birds were usually present throughout the summer months. A juvenile was seen on 14 August 2001, suggesting that birds had bred that year.

At Fernworthy birds were present each breeding season 1997-2002, and were successful 1998-2001, raising two young each year. In 2004 a pair attempted to breed at Burrator, but was unsuccessful.

Since the early 1990s, dispersing birds or non-breeders have been seen at Kennick, Burrator, Tottiford and Venford Reservoirs during the summer months. Two were at Meldon Reservoir in June 1996, but without any suggestion of breeding.

Birds still occasionally occur at reservoirs during the winter, but most tend to be late winter records of birds returning to breeding haunts.

A bird on the River Plym at Cadover Bridge on 22 January 1999 was most unusual, and Dartmoor's only record away from reservoir sights.

RED-NECKED GREBE *P. grisegena*
Very rare winter visitor.
Only one record of a single bird.

> 1987: Burrator, one (presumably the same) seen on 28 January, 4 and 14 of February, and 21 March.

SLAVONIAN GREBE *P. auritus*
Very rare winter visitor.
Three records of single birds.

> 1947: Burrator, 5 and 26 January, one
> 1955: Hennock Reservoir, 7 March, one
> 1963: Burrator, 22 December, one

It is perhaps noteworthy that two of these three records were during particularly hard winters for the region.

BLACK-NECKED GREBE *P. nigricollis*
Very rare winter visitor.
Two records of single birds.

> 1982: Burrator, 14 November, one
> 1987: Burrator, 31 January, and 8 and 14 February, one

MANX SHEARWATER *Puffinus puffinus*
Accidental visitor.
On 21 September 1975 one was found dead at Princetown prison. It had been ringed as a pullus on Skokholm, off the Welsh coast, only twelve days before on 9 September.

This is an extraordinary record, and the circumstances that led to the bird's early demise can only be guessed at.

The adults begin their southward passage from breeding colonies in July, with juveniles following in September. The young birds in particular are very vulnerable to autumn gales, and in Britain a few juveniles are found inland after gales in most autumns (Cramp et al 1977). The Dartmoor bird was no doubt blown off course by adverse weather, and was then perhaps fatally attracted to the lights of the prison at night.

STORM PETREL *Hydrobates pelagicus*
Accidental visitor.
This is another species that is prone to being weather-blown in the autumn gales, although only two have reached the borders of Dartmoor.

> 1932: Near Okehampton, one picked up dead in early November
> 1952: Lustleigh, one picked up alive in late October, but died later.
> > The 1952 record coincided with a "wreck" of Leach's Petrels in the county.

LEACH'S STORM PETREL *Oceanodroma leucorhoa*
Accidental visitor.
D'Urban and Mathew noted this species as a casual visitor to Devon in autumn and winter, when individuals are driven inland by violent gales. This comment holds true for the very few Dartmoor records, which coincide with severe weather.

The first record was of a bird picked up near Rock, Yelverton in a helpless condition on 20 November 1931. Continuous westerly gales at the end of October 1952 provided the most dramatic "wreck" of this species in the South West. Thousands were involved, with at least 63 recorded in Devon. Most were coastal, but Dartmoor had a few, with singles found at Ivybridge on the 25, Tavistock on the 27, and Yelverton on the 30 October. No doubt a great many more went unrecorded.

On 21 December 1989, after more westerly gales, an individual was found dead at Harford. A further bird was found alive at Tavistock after further gales in early November 2000.

GANNET *Morus bassanus*
Accidental visitor.
The 1931 Devon Bird Report contains a note from H.G. Hurrell stating that G. Shillibeer had evidently seen a Gannet at Burrator for a single day during November. The bird was certainly storm driven, and the occurrence may well have been about the same time as the Leach's Storm-Petrel picked up at nearby Yelverton. This is the only Dartmoor record.

CORMORANT *Phalacrocorax carbo*
Regular visitor from the coast, and probable dispersal or passage migrant.
There are very important breeding colonies of this species on the south coast of Devon. Most Dartmoor records refer to birds from these colonies travelling to moorland rivers and reservoirs to fish. The development of the system of artificial reservoirs in the twentieth century no doubt greatly encouraged these inland fishing forays. Pidsley comments on a bird shot at Tavy Cleave, 23 miles from the sea, as if it was a most noteworthy record. Today it would attract little comment.

Birds can be found on almost all rivers, but the largest concentrations are to be found on the reservoirs. Up to five or six are regular at most reservoirs throughout the year, with a bias towards the autumn when juveniles appear with adults. Larger numbers can occur at times and maximum counts are 15 at Hennock on 12 January 1980, 22 at Fernworthy during November 1990, 8 at Meldon on 1 October 1985, and 10 at Burrator in September 1996, and again in the Octobers of 2000 and 2001.

There could well be another reason for the birds being seen over Dartmoor. The British population of Cormorant is thought to be largely non-migratory, but they are known to disperse fairly extensively after the breeding season with some crossing the Channel to northern France, and even wandering to the coasts of Spain and Portugal (Cramp *et al* 1977).

Moore states that migration across Dartmoor has been observed in spring and autumn, but gives no details. Occasionally quite large parties of birds have been noted, passing over at some altitude, in the autumn. The main numbers noted in recent years are:

1979: Brentor, 13 October, 17 flying south
1991: Widecombe, 24 October, 12
1995: River Plym, 21 September, 19 flying south high over Cadover Bridge
1997: River Plym, 18 September, 11 again at Cadover Bridge
1999: Horrabridge, 29 September, 19
2004: Aish Tor, 26 August, 21 flying high westwards in "V" formation

These may of course be unusually large feeding parties going to or from the south coast, but they may also be birds dispersing from colonies elsewhere, perhaps the Welsh coast.

Tree nesting has never been noted in the Dartmoor area, but although most birds leave the reservoirs and rivers at last light to return to the coast to roost, tree roosting has been noted a few times. At Fernworthy Reservoir on 3 January 1977, four birds were noted in a beech tree. Two were watched roosting in shore-side trees at Burrator on 11 February 1981, and up to three were tree roosting on the Dart at Buckfastleigh in 1995. Prior to 2000, birds had been noted roosting in conifers on the West Dart at Huccaby.

BITTERN *Botaurus stellaris*
Very rare winter visitor.
There are six records of single birds.

1946: Sheepstor, during February one frequented a marsh valley above the village
1964: Sticklepath, in January and February one stayed on a small marsh
1969: Grenofen, 21 December, one
1985: Burrator, 24 January, one standing in water by side of road
1987: River Tavy, 24 January, one near Peter Tavy
1998: West Dart, one flushed in early December near the confluence of the Blackabrook

Bitterns tend to move out of their reedbeds in eastern England, at the onset of sub-zero temperatures. This is when many move into the milder south west. The few Dartmoor records tend to occur when there has been hard weather, and something of an influx into the south west generally.

NIGHT HERON *Nycticorax nycticorax*
Very rare vagrant.
There is one record of an adult on the River Meavy near Burrator 17 to 27 May 1999.

SQUACCO HERON *Ardeola ralloides*
Very rare vagrant.
Only one record of a single bird was found acceptable to D'Urban & Mathew and Moore. This was a bird, stated to be nearly an adult, shot near a pool on the Blachford estate, Cornwood, in June or July 1840.

LITTLE EGRET *Egretta garzetta*
Rare vagrant, but possibly soon to become very scarce winter visitor and passage migrant.
The status of this species on Dartmoor is at present difficult to assign.

There were no records at all prior to the late summer influxes that began along the south coast in the late 1980s.

The first was a bird watched on the River Meavy near Yelverton on 31 May 1991. Birds tend to keep well away from the high ground in general, but with ever increasing numbers arriving into the county every year, no doubt feeding pressures have increased, and individuals are turning up more regularly around the edge of the moor. There have been around twenty records between 1991 and the early months of 2003. As would be expected, six of these records were in the late summer to autumn period, reflecting the influx from the continent, and as they were all one-day sightings they were no doubt birds dropping in to feed en route to elsewhere.

Three birds were observed in May, but somewhat unexpectedly ten birds appeared in the coldest months of January to March. Several reservoirs and rivers have had records, but the only places to have had birds in several years are the china clay pool areas of Lee Moor and Portworthy Dam.

There is every indication that more can be expected. In 2003 two took up residence at a Dartmoor location, when birds were present on the River Plym at Cadover Bridge until at least 22 March.

With pairs breeding in Devon, Cornwall, Dorset and Somerset since 1996/98, perhaps it will not be too long before a pair attach themselves to a heronry somewhere on the edge of Dartmoor.

GREY HERON *Ardea cinerea*
Resident breeder.
With birds wandering to even the smallest streams and marshes to feed, the Grey Heron is quite a familiar sight on Dartmoor, where it is frequently seen over-flying the moor between sites. However, many birds come from heronies that are some distance from the moor, and heronies actually on the moor tend to be small and sometimes transient. A short history of known heronies is as follows:

Bellever: A heronry in larch trees here is mentioned by D'Urban and Mathew. There are no details as to the number of nests, and no further information on it until a remark (Dare and Hamilton 1968) that the site had been incorporated into the new plantation area. In the 1960s it was still known locally as "The Cranery". It was not being used at that time and there is no record of when breeding ceased.

Baggator Plantation: The *Devon Bird Report* for 1936 stated that one pair had bred there in larch. There had been a decrease since 1930 when several pairs bred. The decrease was due to many of the trees used being blown down in severe weather. Nothing more is heard of this site, and it is presumed that it did not exist after this date.

Archerton: For many years this was the heronry of the Postbridge area. It was first recorded there in the 1930s, and monitored yearly from 1948 until its abandonment in severe weather in the 1962/63 winter. At 1300 ft. it was the highest heronry in Devon. Numbers fluctuated between one and five pairs, the maximum number being in 1954. The heronry wood, consisting of beeches and firs, was blown down in a storm in January 1990. (Dare 1958; Dare and Hamilton 1968; Dare 1996)

Hexworthy: In 1949 three pairs nested and reared broods successfully by the River Swincombe.

Soussons: One pair bred in 1965-67 and two pairs 1968-73. It was thought to partially replace the Archerton heronry, but there were no further reports of breeding after 1973 (Dare 1996).

Dunnabridge: A pair bred here in 1966. Again, probably a pair displaced from the Archerton site.

Tor Royal: Pidsley mentions a heronry on Dartmoor near Princetown that could well be this site. Birds were still breeding in 1976, when two pairs raised young, but then the predictable happened and the trees were destroyed by a storm. This appears to have been the limiting factory in so many of Dartmoor's high ground heronries. There were no records in 1977, or indeed until 1996, when a pair appeared to nest in a small plantation below the farm. This was, however, a one-off event.

Meavy: In 1979 two pairs were discovered in a riverside location downstream from the village. Two or three pairs then bred annually until 1986, with four in 1983. None nested in 1987 or thereafter. In 1990 four pairs were found breeding at a site downstream from Hoo Meavy. These were presumably birds from the original site. Two to four pairs bred 1991 – 1997, with an increase to seven in 1998. Six were located in 1999, five in 2000. No count was made in 2001 due to the foot & mouth crisis. Breeding continues at this site.

Chagford: Three pairs bred at a site near Chagford in 1977. Nothing more is known from this site. However, on 18 February 2000 four birds were seen circling and calling over Rushford Wood, near Chagford, in such a way as to suggest a probable heronry there. Perhaps this was the 1977 site.

Holne: A rather short-lived heronry existed near Holne from 1979 to 1981. Three pairs bred in the first year, but only two in 1980 and one in 1981.

Venford Reservoir: A pair was first watched here in 1982. Birds bred here up until 1987, when nesting stopped. One or two pairs bred most years, but three pairs were present 1984 to 1985.

Fernworthy Reservoir: A single pair has nested here in the plantation from 1989 to at least 2001.

Harford: A pair bred here in 1993, but for one year only.

Beardown Plantation: In 1994 a heronry was discovered here on the West Dart. At that date it consisted of eight nests, so it may well have existed for some years previously, undetected. It has held between six and eight nests yearly since then, with an increase to ten in 1995. Breeding continues at this site, which is the largest heronry on Dartmoor. The very existence of this heronry was severely threatened in 2004 when much of the plantation was felled.

Horsham Cleave (Wanford Wood): This heronry on the River Bovey was first recorded in 1996. Four to six pairs have been recorded there every year since.

Buckfast: A heronry first recorded here in 1998 still continues. There were six nests in the first year, eight in 1999. The number halved to four in 2000, and again to two in 2001.

As can be seen from the above, Grey Herons have always been heavily reliant on artificially planted conifers for their breeding sites in Dartmoor.

Heronries off the moor no doubt also contribute birds to the numbers that feed on Dartmoor streams and reservoirs. The largest of these is the heronry at Beechwood, Sparkwell that held 24 nests when it was first recorded in 1998, with 25 in 2000. This is now the largest heronry in Devon, and is only a few miles from Dartmoor's southern boundary. Also of significance is the site at Double Waters on the River Tavy, just outside the boundary. Herons have bred here since at least 1986 with nine nests being the maximum in recent years. Other pairs have nested near Okehampton for some years, and at Hemerdon in 1982-3. A site near Tavistock was used 1976 to 1981 with the maximum of eight pairs nesting in 1980.

As mentioned above, feeding birds can occur just about anywhere in suitable habitat, but of the regularly used reservoir sites only Burrator has regularly attracted good numbers. Even here numbers have changed over the years. In the 1960s and 70s double figure counts were regular in winter. Numbers peaked with the count of 18 on 26 December 1969. Since the 1980s numbers have been very much lower with nine being the maximum. Also peak numbers now always occur in late summer, presumably with juveniles in attendance, and not winter as previously. The reasons for this are obscure, perhaps relating to disturbance or fish stocks in the reservoir declining.

PURPLE HERON *A. purpurea*
Very rare vagrant.

In January 1832, a waggoner passing over Whitchurch Down flushed a large bird from the roadside. He struck it down with his whip, and the corpse was eventually passed to Rev. Thomas Johnes, Rector of Bradstone, an amateur ornithologist and collector. He passed the details to Mrs. A.E. Bray in a letter regarding the birds of Dartmoor, which was subsequently published in her three volume work (Bray 1836).

At first glance it is very easy to dismiss this record as a misidentification, because of its strange location and the very unlikely time of year, but Rev. Johnes includes a detailed description of the bird in his letter, complete with measurements, and there is no doubt that it was a Purple Heron, almost certainly an immature just entering its second calendar year. This was the first and only record for Dartmoor.

BLACK STORK *Ciconia nigra*
Very rare vagrant.
Two records of a total of four birds.

1991: Dean Prior, 22 September, three adults flying over.

2001: Marley Head, South Brent, 26 August to 20 September, one.

Amazingly, both records are from the same area on the southern border of Dartmoor.

WHITE STORK *C. ciconia*
Very rare vagrant.
Three or possibly four records of singles.

1949: Dartmoor, 19 to 21 June, one.

1986: Cherry Brook, 6 June, one adult flying south.

1993: Buckfastleigh, 30 April, one flying over the A38.

Near Lustleigh, 1 May, one in the morning.

It is probable that the last two records referred to the same bird.

MUTE SWAN *Cygnus olor*
Rare visitor.
Only four known records.

1964: Peter Tavy, 20 December, seven flying up the Tavy valley.

1976: Burrator, 14 March, a pair present for the day.

1987: Fernworthy, 25 to 27 February, one.

1999: Cadover Bridge Fishing Pools, June to Aug., one. Also noted here later the same year.

It appears incredible to refer to such a well known bird as a "rare visitor", but as the above records show, it keeps well away from the high ground of Dartmoor. They are known throughout the UK to generally keep away from ground over 300 metres (Lack 1986), the reason for this being the acidity of upland streams and reservoirs and the subsequent lack of sufficient vegetation (Sitters 1988).

BEWICK'S SWAN *C. columbianus*
Rare winter visitor.
Four records. A single, a pair, and two family groups.

1963: Burrator, 16 February to 1 March, one.

1968: Fernworthy, 15 December, three.

1970: Burrator, 9 January, two.

1980: Meldon Reservoir, 3 December, two adults and an immature.

Devon is outside of the normal wintering range for this species, and the birds that do arrive are invariably linked with severe weather. The same is true of the very few that have dropped into Dartmoor reservoirs. The only bird that stayed for more than a day was the single at Burrator in the exceptionally severe winter of 1963.

WHOOPER SWAN *C. Cygnus*
Rare winter visitor.
Four records of singles.

    1962: Burrator, 3 to 9 January, one.
    1963: Burrator, 1 to 5 January, one.
    1976: Swincombe, 24 January, one
               Hennock Reservoir, 8 February, one adult.

This is another species linked to severe winter weather, when individuals are driven west to hopefully slightly milder conditions. The record from Swincombe in 1976 is interesting in that the bird involved was attracted to a very small sheet of water. It could possibly have been the same bird that was seen several days later at Hennock, a more suitable sized reservoir.

PINK-FOOTED GOOSE *Anser brachyrhynchus*
Rare feral visitor / escape.
Unfortunately, none of the seven records of this species suggest a wild origin. All records are of singles and several are associated with Canada Geese numbers. Five records are from the Kennick / Trenchford / Tottiford Reservoir complex, and two are from Fernworthy Reservoir. Four of the birds appeared in late June or July.

WHITE-FRONTED GOOSE *A. albifrons*
Very rare winter visitor.
Pidsley mentioned that the species sometimes visited Dartmoor, and that a Mr. Clark had shot a bird out of a group of three at Fox Tor Mire in February 1885.

In more recent times, there have only been three records, all no doubt associated with flocks moving west from regular wintering grounds to the eastward in severe weather.

    1963: Tavistock, / February, 150. This may not have strictly been a Dartmoor record, but as it was in a very hard winter, with a widespread influx into the county of this species, it is worth noting as showing a flocks proximity to the moor.
    1972: Burrator, 2 January, 16.
    1976: Sampford Spiney, 20 November, 16 flying south

GREYLAG GOOSE *A. anser*
Very rare feral visitor, with one record that suggests wild birds.
On 27 December 1981, a group of nine approached from the east over Grenofen, at 100 ft. They circled once, calling, and flew off to the west. This record could well

be of wild birds, but with this species breeding over much of Britain in a feral state, there must still be some doubt. There is very little doubt surrounding other Dartmoor records of three at Lee Moor / Portworthy china clay pits in August to September 1995, with other records during 1997. They were obviously feral birds.

SNOW GOOSE *A. caerulescens*
Very rare feral visitor / escape.
No suggestions of genuine wild birds appear in any of the Dartmoor records.
A blue phase bird was seen with the Canada Geese at Fernworthy Reservoir on 24 October 1992. One was present at Portworthy Dam during August / September 1996, and singles were seen there on 15 May, 15 July and 15 August 1997. Three were seen on 15 September 1997.

CANADA GOOSE *Branta canadensis*
Resident breeder in small numbers. Resident population increased by dispersing birds from elsewhere in autumn.
Canada Geese were introduced into this country in the late seventeenth century, from North America, as ornamental wildfowl. For over the next two centuries birds were confined mainly to large estates and wildfowl collections, and although birds did escape, the small feral breeding colonies that were formed were self contained with no obvious interrelationship. By the mid-twentieth century things had begun to change, and a rapid expansion was under way.

In Devon the first birds were introduced into Shobroke Park in 1949. There was a gradual spread from there into much of mid and east Devon, with birds first reaching Dartmoor in 1955, when two wanderers were at Hennock Reservoir on the 11 April.

Since the late 1970s birds have also spread to sites on southwest Dartmoor, from the breeding population originally on the Lynher Estuary in East Cornwall.

Being largely grazers, Canada Geese take readily to breeding near small areas of water, as long as sufficient grass feeding areas are close by. Not being dependent on aquatic plants, they can use many upland sites that would not be suitable for most other wildfowl due to the acidity of the water (Sitters 1988).

The first recorded successful breeding on Dartmoor was at Hennock in 1964. Thereafter most reservoir sites were colonised, with the first breeding attempts at Burrator in 1966, Fernworthy the same year (Dare and Hamilton 1968) and Meldon in 1981. Most sites held a single pair in any one year, no doubt due to the aggressive territorial nature of the species, food availability and lack of suitable nesting areas. However, two pairs bred at Hennock in 1970, 1980 and 1983, Meldon in 1988 and 1998, and Burrator also in 1998.

Shortly after the initial first successful breeding in the mid 1960s, it became apparent that birds were also taking an interest in nesting sites beside very much smaller areas of water than the main reservoirs. In 1969 and 1972 a pair bred on a small reservoir at Bovey Tracey. On the 3 May 1970 a bird was found sitting on three eggs by the flooded disused clay pit at Redlake, in the centre of the southern moor. 1972 saw a pair attempting to breed on the small reservoir by Yarner Wood. The fishing pools at Cadover Bridge held a successful breeding pair in 1985.

A pair chose the flooded clay pit at Crownhill Down to rear young in 1991.

By the early 1990s the species had increased to such an extent that working clay pits were being used. This activity was centred on the active pits in the Lee Moor / Portworthy area. In 1993 a pair chose to nest on an island in the middle of a pit, and in 1995 five pairs used these same pits but all attempts failed. By 2001 twelve or thirteen pairs were present on 25 April, and five or six pairs bred, raising broods of two to seven.

Small ponds at Throwleigh, Horsham Pond, Luckdon, Brisworthy Pond, North Bovey and the West Webburn valley all held successful pairs between 1990 and 2001, raising between two and seven young, with three or four the norm.

After the breeding season the parents usually stay for some time with the fledged young. It is probable that during this time, adults go through their period of moult in which they are flightless. As they are then very vulnerable they tend to congregate on the larger reservoirs for safety. Immediately after this flightless period there is an influx of moulted birds and presumably juveniles from elsewhere. Numbers swell in August and September, with Portworthy and Fernworthy in particular showing regular high numbers. The maximum count at Portworthy in recent years has been 445 in September 1999, and at Fernworthy 186 on 22 September 1996. Hennock also shows an increase at this time with 146 on 18 August 1995 being the highest count.

Numbers on nearby Bodmin Moor are considerably greater than on Dartmoor, with favoured reservoirs there regularly attracting counts in four figures in recent years. Interestingly maximum counts there are earlier than on Dartmoor, usually in June or July, suggesting a major congregation of moulting birds.

Numbers at Dartmoor sites quickly drop in October, as birds move away from the high ground for the winter months, although Hennock tends to hold birds during the winter better than the other more exposed reservoirs.

BARNACLE GOOSE *B. leucopsis*
Rare feral visitor / escape.

Nothing in the occurrence of the 30 or so Dartmoor birds since 1979 suggests that wild birds have been involved. The fact that the peak months for records are August and September speaks for itself.

Birds quite often attach themselves to post-breeding flocks of Canada Geese. Fernworthy has been the reservoir with the most records followed by Hennock. In recent years the china clay pits of Lee Moor / Portworthy have attracted birds. Most records there are of birds arriving at the peak time of Canada Geese movement, but in 1995 up to four were at Portworthy throughout the year.

Interestingly some of these individuals may not have travelled very far, as birds were known to have bred in captivity in the early 1990s at Cornwood, perhaps on the Blachford estate.

33

BRENT GOOSE *B.bernicla*
Very rare winter visitor or passage migrant.

The only record is of a bird of the dark-bellied race *B. b. bernicla* found grazing in a field by Fernworthy Reservoir on 11 November 1999. It was gone the next day. It had turned up at a time when wintering numbers build up on the Exe Estuary, the only regular wintering site in Devon and Cornwall. Perhaps it became disorientated due to poor visibility and over-shot the Exe.

SHELDUCK *Tadorna tadorna*
Rare passage or moult migrant.

Shelduck breed on coastal estuaries and are rarely seen inland. After their young have fledged, most adults leave and travel to moulting grounds either on the Dutch or German coasts, or possibly Bridgwater Bay.

The return journey from the moulting grounds is gradual and less direct. Many UK birds stay on the German/Dutch moulting grounds until December, then tend to move back in waves to the estuaries of eastern England, moving on west and north from there (Wernham et al 2002). Many do not arrive back in the South West until early spring.

The above needs to be stated so the records of birds on Dartmoor can be explained. Some are very clearly birds on the post-breeding moult migration. The twelve flying NE at about 1000 ft. over Sampford Spiney on 20 June 1977, and the bird at Meldon on 11 June 1978 were obviously in this category. But the records in April and May are a little more puzzling. Are they non-breeders moving away from the estuaries early on their moult migration, or very late birds returning to the estuaries from last winters moulting grounds? The pair at Fernworthy Reservoir on 10 May 1992 can be included in this category, as can the single at Meldon Reservoir on 24 April 1997, the four at Portworthy on 2 May the same year, and the five at Portworthy on 23 April 2001.

Birds returning from moulting grounds no doubt included the bird at Hennock on 12 January 1980, the single at Burrator on 26 February the same year, the bird at Burrator again on 7 November 1981, the bird at Portworthy in October 1995, and the individual that dropped in briefly at Meldon on 2 December 2000.

MANDARIN *Aix galericulata*
Very scarce, but increasing, resident. Stock derived from naturalised introductions and escapes.

The first bird recorded on Dartmoor was a male at Burrator Reservoir during January 1960. It was another 26 years before the next, of a male at Moorhaven near Bittaford, was recorded on 1 and 6 April 1986.

However, between these dates things had been happening to the status of the species in Devon. The 1960 record would rightly have been considered an escape. But by the time of the second record, Mandarin were establishing themselves as breeding birds in the county. In 1974 two pinioned birds were obtained for the wildfowl collection at Saltram House near Plymouth. They later became free-flying, and although often returning to the pond at Saltram for food, became established in suitable breeding habitat on the R. Plym (Sitters 1988). They were first noted at Plymbridge Woods in the winter of 1980/81, and the first breeding was proved in 1984 when two broods were seen. From then on the birds made their way ever further up the Plym and Meavy rivers, breeding at Bickleigh Bridge in 1985. Birds were seen at Burrator October to early December 1989, January 1990, and October and November 1992. A pair were also seen at Meavy Ford on 3 May 1993. Presumably from the same original source, a male reached the china clay pits at Lee Moor on 7 March 1994, and birds have been regular in the spring at the nearby Portworthy Dam from 1996, with breeding confirmed in 1998, when a female was present with three young. In August 2004 up to twelve birds in female/immature or eclipse plumage were on the clay pits at Lee Moor, suggesting a moulting site.

In 1987 five drakes were seen on the Trendlebeare Reserve, river Bovey on the 10 November. Their source was presumably unknown, but from 1990 birds became regular in the area. Birds appeared in Yarner Wood in 1990, and since the species is largely a tree-hole nester, took quickly to the area, with the first nesting attempt in 1994 when a pair used a hole in a beech tree, but the nest was predated by Grey Squirrels. Three pairs were present in 1995, and in 1996 successful breeding was confirmed. In 1998 a pair used a nest box but the outcome was unknown. Four pairs were present during the breeding season of 1999, but successful breeding was not confirmed until the next year, when four or five young were seen on 15 May. Elsewhere in the Bovey Valley, birds were seen on the Parke Estate from 1994, with breeding in 1997 when 6 young were seen. Two pairs were present there in 1999. Four birds, including two males, were seen higher up the river at Neadon Bridge on 4 December 2001.

On the Dart birds also appeared in the late 1980s. A pair was recorded at Buckland Bridge in 1988 and 1989, and in 1992 and 1993 birds were seen at New Bridge. A DNPA survey of the Dart extending 65 km upstream from the River Dart Country Park, and covering both the East and West Dart, found four males between 5 May and 16 June 1993. By 1997 Mandarins were certainly increasing on the Dart, with pairs or singles noted at Holne and Chase Woods, New Bridge, Spitchwick and Hembury Wood. 1998 was the year of the first recorded successful breeding on the Dart, when seven young were seen with a female at Holne Woods on 28 May, and nine young and a female were watched downstream at Buckfast also in May. Unfortunately the nine young there had reduced to two by August. Since then broods have been seen at Buckfast in 1999 and 2000, and at Hembury Wood in 1999-2001. An interesting sight in the recent years has been of one or two birds in eclipse plumage, turning up on a quiet stretch of the West Dart at Huccaby in June or July, presumably to moult. Birds stayed in the area until early September in 2000 (Hibbert 2000). In 2003 and again in 2004 pairs bred at this site.

Elsewhere, away from the three mentioned river systems, a female was present in Okehampton on the R. Ockment on 24 October 1989, a male at Chagford in

March 1996, another noted occasionally at Ashburton in 1997, and a further male at Meldon Reservoir on 16 November 1999.

WIGEON *Anas penelope*
Very scarce winter visitor, and occasional passage migrant.

Birds appear on Dartmoor reservoirs at times in winter. Occasionally a small flock will put in an appearance for a day. The largest number recorded on any reservoir was 79 at Burrator on 29 December 1971. Fernworthy had a flock of 24 present on 3 January 1977. Otherwise wintering birds have been a maximum of eleven, with one or two being the norm. Most favoured localities have been Burrator and Fernworthy, with Hennock having three birds on one occasion, 2 February 1992.

A flock of 32 were seen at Fernworthy on 28 October 1997, and this is the only recorded instance of autumn passage birds at a moorland site.

GADWALL *A. strepera*
Very scarce winter visitor.

D'Urban and Mathew considered Gadwall a very rare duck in Devon, and although numbers in the county have increased considerably since the late nineteenth century, it is not regular at all on Dartmoor.

The first record was of four birds on Burrator on 30 December 1952. Burrator held a single on 12 November 1955. Then there were no records until two were seen again at Burrator on 12 December 1976. One or two were at Burrator in February 1977 and December 1980. Fernworthy had its first record with a male on 13 December 1980, and its second record with a male again on 22 January 1983.

1996 saw the first real multiple numbers on the moor when five were at Meldon Reservoir on 31 January, and seven at Kennick Reservoir on 26 and 30 December. Both of these groups occurred at times of extreme cold when high numbers were present elsewhere in the county. Cold weather was still playing a part in January 1997, when the December 1996 birds at Kennick had been reduced to six on the first of the month, and the largest number to date appeared at Burrator, when up to ten birds were present from 29 January into March. This was also the only instance of birds staying at a Dartmoor site for a protracted period.

A single at Burrator on 21 April 1977 was the only record of a passage bird being seen on the moor.

TEAL *A. crecca*
Winter visitor and passage migrant. Rare breeder.

D'Urban and Mathew found pairs of Teal on Dartmoor bogs in mid-April, and thought that there may have been an occasional nest in the least disturbed places. This comment could still well apply today.

The problem of proving breeding with this species has always been that the broods tend to stay concealed amongst water-side vegetation. This makes matters rather difficult, and could lead to under-recording (Holloway 1996). This is certainly true of Dartmoor, where Teal tend to turn up in the breeding season on small pools in the blanket bog, as well as in valley bottoms and on certain reservoirs. Thus birds have been reported in the breeding seasons at 22 sites since 1930, but breeding or attempted breeding has only been confirmed at four of these.

Confirmed breeding was first recorded at Burrator in 1930 and again in 1932, at Fernworthy in 1948, 1999, and 2000, at Manga Brook Head in 1993, and on Okement Hill in 2002. Young were seen in the 1930s Burrator records and the 1948 Fernworthy breeding. At Manga Brook Head in 1993 a nest with seven eggs was located 29 April and a female with at least one young was still present on 31 May. This was later found abandoned. As the first eggs are usually laid in mid-April, nearly all of the remaining breeding season records of birds in suitable habitat could have referred to breeding pairs.

Breeding was suspected or at least thought possible at Tavy Cleave in 1935, Erme Valley in 1938, Knattabarrow Pool in 1942 and 2000, Fernworthy in 1952, 1953 and 2001, Fox Tor Mire in 1979, Sandy Hole Pass (E. Dart) in 1981 and 2002, Yealm Plains in 1981, Swincombe Head in 1982, Red Lake Mire also in 1982, Upper Wallabrook in 1988, Hew Down in 1989, Raybarrow Pool in 1992 and 1999, between Gt. Varracombe and Teignhead Farm in 1993, Gallaven Mire also in 1993 and 1999, Whitemoor Marsh in 1994, Blackabrook (Princetown) also in 1994, Cheriton Coombe in 1998, Taw Head in 2000, and Lee Moor in 2002.

It can be seen that many of the above cases of suspected breeding took place in the area of the North Teign River and its tributaries. The reason for this grouping is at present unclear. Is it that this area holds a particularly favourable habitat for the species, or does it get particularly good coverage in most breeding seasons? The other grouping is on the southern moor around the upper reaches of the Yealm, Erme and Swincombe. All these sites were located during fieldwork for the Devon Tetrad Atlas (Sitters 1988).

With this number of confirmed and suspected sites, and taking the difficulty of proving breeding into consideration, it must be concluded that although rare, Teal must be a more regular breeder on the moor than records would suggest.

After breeding, birds may well stay around in suitable feeding areas, but it is usually late August or September before passage birds begin to arrive. Numbers increase gradually in October, and then merge with the wintering population. Burrator and Fernworthy tend to be the only reservoirs to regularly hold wintering numbers.

Teal in winter are very dependant on weather conditions. They feed in shallows and wet mud, and as these are the very areas to freeze first in cold spells, they must move location as soon as this occurs (Wernham et al 2002). Hence, numbers on Dartmoor can vary from year to year depending on local conditions, and conditions further north and east. In the 1960s and 1970s numbers both at Burrator and Fernworthy would regularly average about 35/50, but in more recent years numbers have fallen to about 15/25. Exceptional counts at Burrator were 60 on 14 November 1971, and 90 on 25 January 1976. At Fernworthy maximum counts have been 49 on 1 February 1980, and 53 on 6 December 1998.

It is possible that the autumn/wintering population may be rather higher than proved, as parties can be easily overlooked on small wetlands and private ponds. Instances of this were 50 on ponds at Whiddon Down 8 October 2002, and the 30 on flood water pools at Dunnabridge on 20 December 1968, with 17 at the same locality 22 November 2001.

GREEN-WINGED TEAL *A. carolinensis*
Very rare vagrant.
Three records of single males.
>   1969: Burrator, 12 to 27 April, one.
>   1974: Burrator, 15 to 18 December, one.
>   1997: Knattabarrow Pool, 6 March, one.

MALLARD *A. platyrhynchos*
Resident breeder, passage migrant and winter visitor.
The well-known Mallard is a most successful species. Although it is largely a lowland breeder, it is found quite widely along Dartmoor rivers and streams, around reservoirs, and in valley bogs. The Waterways Birds Survey on the R. Meavy in 1992 produced 14 pairs, with 10 in 1993, and 11 in 1994. Pairs bred on all five Dartmoor reservoirs in 1988, and indeed probably do most years. The major reservoir sites are Fernworthy, where between two and five pairs breed regularly, and Burrator where two to four pairs breed. Other noteworthy breeding areas have been on the Blackabrook (Princetown) where five pairs raised five or six young each in 1992, and on the West Dart, where three broods were raised at Huccaby in 2000. Individual pairs will breed even some distance from a major stream as a pair did at Vitifer in 1995, raising seven young.

Mallard are thought to be principally a summer visitor to the streams and bogs of the Postbridge area, arriving from mid-March. 15-20 pairs have been reported nesting until the mid 1960s (Dare and Hamilton 1968), although this number was thought to have reduced to about 10 by the 1990s (Dare 1996).

Post-breeding birds tend to congregate on reservoirs from late summer, and it is at this time that Fernworthy gets its highest numbers, with the maximum of 33 in August 1998 and the same number in July 2001. At Burrator the position is different, with the highest counts always being in the winter months of November, December and January. 148 were present here on 4 January 1972, and 123 on 26 December 1959, although in more recent years winter counts at Burrator have rarely exceeded 60. Hennock Reservoirs have had maximum counts of 92 in September 1998, and 78 in November of the same year.

During the autumn and winter, birds may choose to feed and roost in different localities. Numbers of up to 19 were seen to fly in at dusk to roost at Venford Reservoir in October 2000 and November 2001. Conversely, at Burrator on 22 November 1992, 30 birds that had been present all afternoon left the reservoir at last light to roost elsewhere.

Finally, a note of caution must be sounded regarding the true wildness of many of our birds. Many could well originate from captive or feral stock. Birds on the West Dart at Huccaby were believed to have spread from the feral birds at Dartmeet, and the flock of about 120 that suddenly appeared on clay pits at Lee Moor on 22 September 2000 were later found to be from a nearby farmer's duck ponds. With numbers like this dispersing into the countryside, the wild ancestry of the majority of birds must be open to doubt.

PINTAIL *A. acuta*
Rare winter visitor and occasional passage migrant.
This duck has never been at all regular on Dartmoor. D'Urban and Mathew knew it as a rather rare visitor to Devon with no records from Dartmoor. It has occurred in eleven years since 1934, but all have been single day records, and most have been in direct response to severe weather conditions elsewhere.

Burrator is the most favoured location, with singles on 20 March 1934, 11 November 1962, and 11 February 1981. A pair was present on 18 December 1980 and 9 February 1991. Small parties of four were seen on 4 January 1963 and 7 February 1976, and a larger group of ten occurred on 3 February 1954.

Burrator's monopoly of records ended in 1994 when a female was seen in the unusual locality of Tavy Cleave on 1 April, and a female visited the clay pits of Portworthy during October. A party of five dropped in to Meldon Reservoir on 30 January 1996, and a bird was at Parke, Bovey Tracey on 15 February the same year. The latest record is of a male, again at Portworthy, on 14 November 1998.

As can be seen from the above, passage birds do occur at times, but the overwhelming number of birds have been in the cold months of January and February.

GARGANEY *A. querquedula*
Very rare spring passage migrant.
Just two records of single birds.
1969: Burrator, 9 March, one.
1984: Burrator, 7 May, a male.

SHOVELER *A. clypeata*
Rare winter visitor and passage migrant.
This species, too, appears occasionally on Dartmoor reservoirs only to be gone again the next day. Although there are very few records, birds have occurred in six months, with most in March, suggesting an early spring passage.
1949: Burrator, 12 December, a female with Mallard.

1952: Fernworthy, 1 May, a pair.
1959: Avon Reservoir, 19 March, a pair.
1962: Burrator, 3 January, five.
1967: Hennock Reservoirs, 19 March, two.
1975: Fernworthy, 2 February, four.
1981: Burrator, 18 February, a male.
       Burrator, 29 October, a female
1997: Burrator, 10 March, a pair.
2000: Portworthy, 15 October, three

POCHARD *Aythya ferina*
Winter visitor and passage migrant.

Before the establishment of reservoirs on Dartmoor, the occurrence of this diving duck was unknown. It is a species that needs large areas of fresh water as winter habitat, and although moorland reservoirs are favoured by small numbers, the shallow, less acid lowland waters always hold the bulk of the county's wintering population.

Although the numbers have always been fairly low, two reservoirs in particular have regularly attracted birds. The reservoirs at Hennock have held birds almost every winter since at least the 1930s. Up until the late 1970s counts of 30 to 50 were quite regular, with maximum counts of 100 on 29 December 1961, 80 in the second winter period of 1975, and 62 on 18 February 1959. Since the late 1970s, numbers have been generally reduced, with 15 to 25 per winter being the norm, and the maximum count being 29 in February 1989. Numbers of wintering birds nationally are known to have decreased in the late 1970s and 1980s (Wernham *et al* 2002), so perhaps the local decline at Hennock may have been linked with this countrywide reduction, although with such relatively small numbers involved, perhaps a more local reason is more likely. The maximum numbers were almost certainly linked with severe weather conditions.

At Burrator, the second regular reservoir site, numbers have never approached the figures at Hennock. Since the first counts here in 1929, numbers have rarely exceeded 30, although 42 were counted on 5 February 1959 and 40 during January and February 1947. The 1947 high counts were at a time of very severe weather conditions. Numbers here showed a slight reduction since the late 1970s, but not as marked as at Hennock. Since about 1990 numbers have drastically reduced, with the species being absent some years. The reason for this is not fully known, but it could be linked with the completion of Roadford Reservoir about this time, which, being a lowland water, would be more attractive to Pochard.

Fernworthy has only ever irregularly held a few birds. But in 1976 it broke the rules with 43 present on 2 January and 30 on 26 December. On 16 January 1974 34 were also here. The only other site to hold double figures is Portworthy where 32 were on the clay pits in January 1996, with 13 in January 1997.

Dartmoor's population of Pochard is basically a wintering one with the highest numbers occurring in January and February, and falling rapidly thereafter. However, a few migrants pass through some years in October, with the occasional record from mid-September. The highest October count is 13 at Hennock in 1993.

RING-NECKED DUCK *A. collaris*
Rare vagrant.
The first Dartmoor record of this North American duck was an adult male that appeared at Burrator on 2 January 1972 and stayed until 29. This bird arrived with Tufted Duck and Pochard after a period of strong easterly winds. It had presumably arrived some time earlier from across the Atlantic, and moved westward with other *Aythya* species, perhaps from eastern England or the continent (Smaldon 1982).

In December 1980 there was a mini-invasion of birds into Devon and Cornwall. Birds moved from place to place quite rapidly, so it was never satisfactorily established just how many were involved. Dartmoor had records of four or five individuals.

Most of the records were from Burrator, where a female was present from 7 to 9 December, and again from 16 to the end of the year. Two immature males were found there on 8 December and remained until 10. What was presumably one of them was recorded again on 27. An adult male was present on 16 and remained with the female until the end of the year. Both the adult male and the female remained into 1981, and were last recorded on 4 April. At Fernworthy an adult male was present on 9 December, which may or may not have been the adult male later at Burrator.

On 5 December 1998 an adult male was found at Burrator with Tufted Duck and was last seen on 23 March 1999. It then became something of a regular, returning with the Tufties on 23 December 1999, and staying until 1 April 2000, and again returning in early November and staying until 3 March 2001. It returned on the 6 December 2001 but only stayed until 15 December. It was later relocated at Lopwell Dam, off the moor, where it stayed for the winter. It did not return in November or December 2002.

FERRUGINOUS DUCK *A. nyroca*
Very rare vagrant or escape.
Two were present on Burrator on 23 November 1961, and what was thought to be one of these birds was seen again there on 3 January 1962. The only other records refer to a tame male that was present on the Dart at Buckfastleigh from 1 May to 1 September 2000, and again on 24 July 2001.

TUFTED DUCK *A. fuligula*
Winter visitor and passage migrant. Very rare breeder.
The Tufted Duck has increased nationally as a breeding species in recent years. Between the two *Breeding Atlases* of 1968-72 and 1988-91, the breeding population increased by 15% and included a range extension into southwest England (Wernham *et al* 2002). However, it is very much a lowland duck, and as breeding habitat, the

waters of Dartmoor are not much favoured.

Breeding on the moor has only been confirmed in the last few years, although attempts may have been made considerably earlier, as at Burrator in 1932 (Smaldon 1982). In fact the first confirmed breeding was not on a moorland reservoir, but on the clay pits at Portworthy in 1992. Breeding again took place in 1994, when two pairs raised one and four young at nearby Lee Moor clay pits. Four young were raised at Portworthy again in 1995.

Birds are normally very few at times of passage, but wintering numbers tend to build through November and December, to peak in January/February. Over the years three reservoirs have regularly attracted small wintering flocks.

The most favoured locality has always been Burrator, with wintering birds in most years since at least the late 1920s. Numbers during the first half of the twentieth century were small, but there was an increase from the late 1960s. The maximum counts there have been 47 in February 1998, 40 on 31 January 1979, 38 in January 1996, 32 in December 1995, and 30 on 9 January 1972. There have been another twelve counts of 20 to 29, all since 1975.

Hennock reservoirs have held birds in the winter since at least the 1940s. Numbers generally have been a little less than at Burrator, although on two occasions it has produced the highest number at any Dartmoor site. This involved 50 being present on 29 December 1961 and also on 8 February 1970. Another high count was 30 on 17 February 1962. There have been a further eight counts of 20 to 29, half of which were in the mid-1970s.

Although it has had wintering numbers since at least the 1940s, Fernworthy comes a rather poor third as a wintering locality. The maximum counts here have been 22 on 25 October 1979 and 14 January 1974. No other counts have exceeded thirteen.

Numbers elsewhere are very infrequent and usually limited to ones and twos, although since the mid-1990s, birds have quite regularly occurred at Portworthy in the winter, with an unusually large count of 25 in January 1996.

The most productive years for wintering birds has been the period of 1970/76, with especially good numbers on moorland locations in January/February 1970.

SCAUP *A. marila*
Rare winter visitor.
This duck, which is largely maritime during the winter around UK coasts, has been found twelve times at Dartmoor sites. The majority of records have been in February or December.

The first record was of a first winter male picked up at Bovey Tracey on 11 February 1954. Burrator has found favour with birds on five occasions, with a male being present 7 February to early April 1976, a pair on 1 December 1985, a female 6 to 22 November 1990, a female on 27 July 1993, an a pair on 24 March 1996. Fernworthy had a female present on 8 December 1987, and another female 19 January to 11 April 1992.

All other sites had birds on only one year. One female was on a flooded clay pit at

Cadover Bridge on 27 December 1978, a female was at Venford Reservoir 15 to 27 December 1991, a first winter male at Trenchford 23 January and 25 February 1995, and a female at Lee Moor clay pits 15 February to 1 March 1996.

As most adult females are known to moult on the breeding grounds after the nesting season, (Cramp *et al* 1977), the July record from Burrator in 1993 is rather puzzling. It is possible that this bird was still immature, not in condition for breeding, and had stayed in winter and passage areas, rather than returning to breeding grounds. What was almost certainly the same bird was seen at Portworthy in August.

LONG-TAILED DUCK *Clangula hyemalis*
Very rare winter visitor.
This bird has occurred at Burrator in three winters. At least two of these occurrences were linked with severe winters.

> 1946: 29 December, two immatures. They were still present on 5 January 1947, with one remaining until the reservoir froze over shortly after 26 January.
> 1961: 9 November, one female stayed for a fortnight.
> 1978: 2 December, one female.

COMMON SCOTER *Melanitta nigra*
Very rare passage migrant.
This species has been recorded on Dartmoor reservoirs in three years.

> 1981: Trenchford Reservoir, 8 July, three males and two females. The time of year strongly suggested that these were grounded overland nocturnal migrants en route to moulting grounds.
> 1982: Meldon Reservoir, 14 to 17 April, a female.
> 1989: Burrator, 8 November, a male with six Tufted Ducks. They arrived during a period of gales.

GOLDENEYE *Bucephala clangula*
Very scare winter visitor.
This is a species that has occurred regularly on Dartmoor reservoirs since their completion, but almost always in very small numbers. The first bird recorded was as early as 1910, when one was shot at Burrator (Smaldon 1982).

Since the early 1930s birds have been seen on one or more reservoirs almost annually, with Burrator being the most favoured site. There, one to three birds have been the norm, with four on a few occasions, and five in January to March 1981 and February 1982. A very unusual record was of twelve present on 24 December 1968. Numbers have decreased greatly at Burrator since the early 1990s, perhaps due to the newly opened Roadford Reservoir offering more acceptable wintering habitat.

Records from Hennock reservoirs are somewhat intermittent, and usually of only one or two birds. Unusually, eight were there on 5 December 1974. Fernworthy, likewise, only occasionally holds birds, with the maximum being five on 8 November

1973. Venford is something of an anomaly. It very rarely has a Goldeneye at all, and yet it has recorded the biggest flock to date on Dartmoor. This was the 15 present on 13 and 22 January 1985. This was a time of severe winter conditions over much of southern England and the near continent, and there were large influxes of many species. Avon Dam, Meldon Reservoir and the clay pits at Portworthy, have all held one or two on a few occasions.

There is little sign of any autumn passage on Dartmoor, with the first wintering birds arriving in mid-to late November. By January numbers are at their peak, and there are fewer records in February, no doubt due to birds beginning to return to breeding areas. Interestingly Burrator is the only site that has had records of birds in March. Some years these have been wintering birds present since January, but in several years birds that have not previously wintered have been observed, obviously spring migrants. In 1986 three birds stayed on here until 2 April.

Unusual records away from reservoirs have been the single bird seen on the River Dart at Spitchwick on 20 November 1982, and the pair on the West Dart near Hexworthy on 13 December 1990.

SMEW *Mergellus albellus*
Rare winter visitor.
When winters are at their hardest in eastern England and the Low Countries, then Devon may see the arrival of small numbers of Smew, and Dartmoor reservoirs may just possibly see a bird or two. Most records coincide with the very coldest of conditions in January and February.

Burrator has had birds, usually redheads, on 28 November 1947, 29 January to 7 February 1984, 20 January to 19 February 1985, a male 16 to 21 February 1986, 7 and 8 February 1987, with two redheads present on 27 February 1991.

Hennock hosted a redhead in January and February 1976, and a male and a redhead on 8 January 1979. A male was at Fernworthy on 21 February 1954, and a male again there on 12 February 1976. The most recent record is of a redhead at Venford Reservoir on 10 February 1992.

RED-BREASTED MERGANSER *Mergus serrator*
Very rare passage migrant and winter visitor.
Although regular as a winter visitor to certain Devon estuaries, Red-Breasted Mergansers are very rare inland. It is interesting that a number of the Dartmoor records would appear to refer to overland passage migrants rather than wintering birds. Dartmoor's nine records are as follows:

1952: Venford Reservoir, 22 March, one.
1953: Near Tavistock, 24 October, one resting on a small reservoir. This record may not have referred to a Dartmoor location, but is given here on the assumption that it did.
1968: Burrator, 12 November, one.

44

1981: Burrator, 6 December, one.

1987: Fernworthy, 4 to 9 December, one. Stated in the Devon Bird Report to have been a drake in eclipse plumage. Given the time of year more likely to have been an immature drake showing early signs of moulting into adult type plumage.

1988: Fernworthy, 6 March, two.

1989: Avon Head Pool, 18 April, one.

2002: Venford Reservoir, 11 and 12 November, a brownhead came into roost with Goosanders.
Burrator, 3 December, one brownhead in afternoon with Goosanders. No doubt the same bird seen earlier at Venford. This individual stayed to winter with local Goosanders, and was seen again at Burrator on 11 and 17 February 2003.

2004: Burrator, 31 January, 11 and 21 February, one male with the Goosanders.

GOOSANDER *M. merganser*
Resident breeder in small numbers, winter visitor and passage migrant.

Goosanders first appeared on Dartmoor reservoirs in the 1930s. The two at Burrator 14 to the 21 February 1931 were probably the first. They were then a scarce to rare winter visitor, and their status remained such for the next fifty years.

The early winter months of 1979 were very hard, and there was a large influx of Goosanders, as well as other waterfowl, into the county from the Low Countries. Increased numbers were noted at Dartmoor locations. In 1980 the first breeding was recorded in Devon on the lower Dart at Staverton. Colonisation began slowly but by 1985 the first Dartmoor breeders were recorded with females with downy young being seen at Dartmeet on 9 May, and Hembury Woods on 11 May.

The River Dart continued to be the main location of breeding pairs on the moor. The Hembury Woods area had confirmed breeding again in 2000, with birds in probable breeding habitat additionally in 1988, 1989 and 1991. Dartmeet had a further confirmed breeding in 1999, as well as birds present during the breeding season in 1997. Others have been found in the stretch between Holne Chase and New Bridge, where breeding was confirmed in 1987 and pairs were located in 1988, 1989, 1992, 1997 and 1998. Further down the Dart at Buckfast and Buckfastleigh breeding was confirmed in 1997, 1998, 1999 and 2001. It is probable that more than one pair was successful in 1999 and 2001. Birds have been noted there most years since 1986. On the East Dart, above Dartmeet, a pair were successful at Bellever in 1993. On the West Dart breeding was recorded in the Huccaby area in 1992, 1995, 2000 and 2001. Two pairs bred here in 2000. The area around Dunnabridge on the West Dart has proved favourable to the species in five years since 1988, but breeding has not been proved. Occasional summer sightings have also occurred at Postbridge, Two Bridges, Lucky Tor, River Dart Country Park and in the Double Dart Gorge.

Male Goosander

The only full survey of the Dart took place on behalf of the Dartmoor National Park Authority in 1993. It covered 65 km of the Dart above the River Dart Country Park, and took in both the East and West Dart. It found six broods of young with another two possible pairs present. In 1996 four broods were located on the Dart with a further five other pairs, although sites were not fully recorded.

On other river systems breeding pairs started to show in the early 1990s. Breeding was confirmed on the Tavy at Double Waters in 1991 and again in 1996, and the Meavy had its first confirmed breeding in 1999, although birds had been present two years earlier.

Since the mid-1980s records have increased at all reservoir sites during the spring and summer months, but only Burrator can claim proved breeding, with broods in 1988 and 1999 and possibly in 1987. At this and other reservoirs much of the increase may be due to fledged juveniles dispersed from their natal river sites. This would certainly seem to be the case at Venford, where summer numbers rose rapidly from about 1994, with 20+ present in late summer some later years.

Nesting sites are known to be holes, often in trees, rabbit holes and the like, but the only three recorded Dartmoor sites were all trees. At Yarner Wood in 1993 a pair nested in a tree-located owl nest box. Six eggs were laid but they were subsequently predated, probably by a Grey Squirrel. A bird was unintentionally flushed from a tree hole at Grenofen on the River Walkham in 1997. Near Dartmeet in 1999 a nest was located in an ivy covered tree.

When females are incubating, males leave the breeding grounds and depart on a moult migration. It is now known that they travel to northern Norway, where they moult and become flightless in four large fjords (Wernham *el al* 2002). Most Dartmoor males disappear from about mid-May, presumably to join the thousands assembling in Norway. However, occasionally a male will stay on Dartmoor waters. Such was the case with a male that was present at Burrator throughout May and until at least 13 June 1995, when it was in eclipse plumage. Another male entering eclipse plumage, but not yet flightless, was seen also at Burrator on 19 July 1993.

An interesting observation that has been made recently is that of communal roosting during the breeding season. At Venford on 5 April 2002 14 birds, including six adult males came into roost at dusk. A month later on 7 May, 16 birds, including eight males, came into roost. By the 17 May 11 birds were still roosting, although by now only one was a male. Presumably the other males had departed for their moulting grounds. As the number of males present in April and in early May suggested established pairs, presumably breeding, do females leave eggs or young in the nest for at least some portion of the night unattended, or could these birds be non-breeders?

Most of the birds present on the moor in the autumn and even early winter are brownheads, either females or young of the year. Adult males begin to drift back from Norway in November (Werner *et al* 2002) and this is reflected in the low numbers at the Burrator, and particularly Venford roosts, until the latter part of November or even December. Resident bird numbers are augmented by winter visitors from elsewhere in December and January, largely dependent on the severity of the winter.

During the winter, birds feed by day on rivers, either on or off the moor, but return to roost most nights at established reservoir sites. Numbers present at roosts vary from night to night, perhaps because birds vary their roosting site, depending on their daytime fishing area, or perhaps some stay on rivers overnight. Birds are known to come into roost very late, contrary to certain statements in the literature, some probably after dark. Counts of roosting birds at first light almost invariably give a higher total than counts the previous evening. Brief details of individual roosting sites are as follows:

Burrator: Roost known here from about 1979. Roosting numbers reached their peak here in the 1980s with high counts of 25 including eleven adult males on 20 January 1982, 28 with two adult males on 2 February 1985, and 43 with eight adult males on 19 February 1987. Since the 1990s numbers have declined generally with high figures only occasional: 24 with five adult males in January 1996, and 26 in December 1999 have been the maximum, but counts just into double figures have been more the norm.

Venford: Roosting noted here since the mid-1980s possibly earlier. Proximity of the Dart, the main Dartmoor Goosander river, has always meant that good numbers are in the area, and thus the roost has been regular, with high numbers. Since 1990 by far the highest counts of any roosting site have been achieved here. On the 18 November 1990 43 birds came into roost including 10 adult males. This total was eclipsed on 12 December 1993, when 55 roosted, including 12 males. This count was on a day of heavy rain and southwest gales, and is the highest count of any Dartmoor site to date. Since then many counts over 30 have been made, with 42, including 14 males, on 17 December 1998, and 40, with 13 males a month later on 21 January 1999.

Fernworthy: Never a great reservoir for Goosanders. Numbers, either diurnal or roosting have always been intermittent and low. Most years fewer than 5 have been present on occasions, with the highest counts being nine in December 1996, and 16 in December 2001.

Hennock Reservoirs: Numbers generally are low. At times, higher numbers resulted from feeding or loafing birds during the day rather than night roosts. The highest number recorded was 21, including 12 males, in December 1990. Eighteen were counted in December 1994 and February 1995. In fact the period from 1990 to 1995 provided all the high counts here. Counts before and after that period rarely exceeded five.

Meldon: Records are intermittent, but the count of 21, including 10 males, at 7.20 am 18 Febuary 2002, shows that, at least on occasions, a night roost does occur. Other high counts have been 12 on 23 January 1994, and 10 on 19 February 1987.

Avon Dam: Since 1963 there have only been seven records from here. As it is a little difficult to get to, this lack of records may not indicate a lack of birds. Nine, with one male, were present on 6 January 1980, and one intrepid fellow, seeing three birds here at dusk on 17 December 1994, walked out to the reservoir again at dawn the next morning to find the number had increased to seven, including one male. They all flew off south over the dam shortly after dawn.

RUDDY DUCK *Oxyura jamaicensis*
Very rare winter visitor.
On 20 February 2004 a female Ruddy Duck dropped onto the the pool at Foggintor Quarries, thus giving Dartmoor a new species. The bird had perhaps been forced down by the weather, as it was not present the next day.

HONEY BUZZARD *Pernis apivorus*
Very rare passage migrant.
This species has always been rare in Devon and especially so on Dartmoor.

D'Urban and Mathew knew of only one Dartmoor record, that of a bird trapped at an unmentioned locality in 1848. The transactions of the Devonshire Association for 1907 contain a note regarding a black morph juvenile that was taken by a game-keeper on the south side of Dartmoor on 20 September 1904.

There then appeared to be no Dartmoor records until a bird in 1971. Although the species bred in the county from 1979 to 1996, this did not signal any significant rise in Dartmoor sightings. It is not impossible that a pair could take up territory in Dartmoor woodland in the future. Records, all of singles, are as follows. The month of August has 50% of the records.

1971: Lydford Forest, 8 May, one.

1977: Vitifer, 16 April, one. An early date.

1980: Burrator, 2 September, one flying SE.

1991: Burrator, 7 August, one.

1994: Burrator, 15 May, one light phase bird going E.
Avon dam, 12 August, one going W.

1996: Roborough Down, 9 June, one put out of its roost early morning by corvids flew strongly E.
Hennock reservoir, 27 August, one.

1997: Trenchford Reservoir, 18 August, one.

1999: Near Two Bridges, 19 August, one drifted SW towards Princetown. It looked like an adult female.

BLACK KITE *Milvus migrans*
Very rare vagrant.

The only acceptable Dartmoor record is of a bird at Sampford Spiney on 24 March 1989.

RED KITE *M. milvus*
Rare passage migrant, with birds becoming semi-resident on occasions.

The early history of this species on Dartmoor is somewhat confusing.

Red Kites were well known and protected as scavengers in medieval Britain.

As sanitation improved in the seventeenth and eighteenth centuries, Kites were no longer needed in urban areas and perhaps now being seen as something of a nuisance, persecution began (Holloway 1996). This persecution also took hold in rural areas, and the Kite, well used to being near man and showing little fear, was an easy target. The decline was rapid and in Devon it would appear that by the early nineteenth century numbers were at a very low ebb.

There are several comments in the literature confirming this. The Rev. Thomas Johnes, writing to Mrs. Bray in the early 1830s mentioned that forty or fifty years previously the Red Kite had been common in the Dartmoor area, but was then very rare and that he had never seen a live specimen (Bray 1836). D'Urban and Mathew stated that it was said to have been common around Dartmoor in the eighteenth century. However, they felt that it had never been as common in Devon as in counties further east, and noted that Col. Montagu, in his twelve year residence in Devon only heard of the Red Kite once, despite having regular correspondence with Dr. A. G. Tucker of Ashburton, a fellow enthusiast and resident in an area close to the moor where Kites were allegedly present.

These sources also quote a somewhat contrary opinion from E.H. Rodd, author of *Birds of Cornwall* (1880), who spent his childhood in Buckfastleigh. Rodd asserted that no other hawk was better known in the large woodlands of the area. He gave a breeding site as Kings Wood near Holne Chase. The period would have been 1820/30. This record is also given in Pidsley. This site presumably is the Kings Wood directly west of Buckfastleigh, rather than near Holne Chase, as (Holloway 1996) gives nearby Dean Wood as a breeding site in 1855, but gives no reference as to source. Pidsley also cites evidence that by 1838 the Red Kite was almost confined to Dartmoor in Devon. D'Urban and Mathew continue by giving details of a bird caught at Trowlesworthy Warren rabbit warrens in 1831, and an adult female trapped there on 17 September 1851, a pair having been seen the previous June. A pair was also shot in Yarner Wood, presumably also in the mid-nineteenth century. One was seen near Chagford in May 1890, but the last known breeding in the county occurred near Dartmeet in 1912 and 1913 (Moore 1969).

It should be pointed out that D'Urban and Mathew were not too confident that all records of Red Kites from early in the nineteenth century, and before, actually did refer to Kites. Rural names for birds of prey tended to get mixed at times, and some references could have referred to other species.

In more recent years the Red Kite has been mainly a rather rare passage migrant on Dartmoor, not occurring every year. Birds have been seen in most months, but by far the majority of records have been in spring, especially in April. Autumn records begin in early October, and continue through November, with two records in December. Rather surprisingly, November has contributed over half of all autumn records.

The habitat favoured by Red Kites on Dartmoor in recent years is largely the moor-edge and wooded valleys that descend from the moor into the agricultural "in country". Many records are from the areas around the southern perimeter of the moor, particularly the valleys of the Yealm, Erme and Avon. Likewise the eastern moor edge area from Chagford to Bovey Tracey has had several sightings. Rather fewer records have come from the Burrator area and the Okehampton / Sticklepath / South Zeal areas. There have been six records from the semi-agricultural mid-moor area of Princetown / Two Bridges / Postbridge, but only two from central moorland locations on the north moor, where a bird was hunting Cowsic Head are on 23 June 1962, and another was reported over Fur Tor on 15 July 1996.

There have been two occurrences of birds remaining for long periods on the western edge of the moor. Both involved birds mainly seen in the Cox Tor area, although they foraged widely over large areas of the western moor.

In 1962 a bird was seen on Cox Tor on 25 November. It remained in the general area until the September of 1963, and was then seen again from 29 December 1963 and throughout 1964. A second bird was seen on three occasions in 1964. One remained in the area until the 19 March 1965, when it decided that it had had enough of a good thing, and finally moved on.

A similar pattern occurred in 1969 when a bird was seen at Sourton Tors on 13 October, and then at Cox Tor on 2 November, where it remained until the end of the year. It was joined by a second bird on 4 November, and this bird too remained until the end of the year. A third bird was seen on 25 November. Two birds remained throughout 1970, ranging widely over the west side of the moor, and one bird stayed until at least July 1972, with a second seen during February and March 1972.

The re-introduction scheme for Red Kite in England and Scotland has brought at least one wanderer to Dartmoor. A bird flushed from its roosting tree at Routrundle (Walkham Valley) early morning of 2 May 2000 had a yellow wing tag, proving it to be a Scottish-bred bird. As these re-introduction schemes are doing so well, no doubt other birds will appear in years to come and possibly stay to breed.

WHITE-TAILED EAGLE *Haliaeetus albicilla*
Very rare visitor. May formerly have bred.
On the Dewerstone, high above the wooded valley of the River Plym, is a crag formerly known as Eagle Rock (Hemery 1983). This is the site where White-tailed Eagles were rumoured to have bred in past times. D'Urban and Mathew and others have repeated this story, but the facts of the matter appear to have been lost in the mists of time. If true, perhaps the site was used so long ago that it became something of a folk legend. Although the link is a little tenuous, an interesting fact was pointed out by H.G. Hurrell in the *Transactions of the Devonshire Association* for 1966. The early medieval masons working on Exeter Cathedral had carved the eagles in the stone work with the wedge shaped tails of White-tailed Eagles. Perhaps our early ancestors were well used to seeing the White-tailed Eagle over southwest Dartmoor.

Whether it ever bred must be open to some doubt, but the species has occurred several times on the moor. One of the later sightings, of a full adult was made by the grandfather of the present writer in 1943. The bird flew north over Wigford Down, an area adjacent to the legendary Dewerstone breeding site.

The full records for Dartmoor are as follows:

1832: In the summer a bird was seen frequently by members of the hunt. What was probably the same bird was shot near Kingsbridge in October that year.

1834: One shot near Bridestowe.

1891: One reported on Dartmoor in May and June.

1934: An eagle, probably of this species, was seen during February by numerous people in the Deancombe area of Burrator.

1936: A first or second year bird was seen in February and March east of Princetown. It was first reported on 15 February. Fox Tor Mire and the Swincombe Reservoir had a particular attraction for it, but it also ranged widely to the southeast, and north towards Laughter Tor. It was last seen on 14 March.

1938: A bird, perhaps the same as the 1936 individual, was seen in the same area of the moor three times in June.

1943: An adult was seen flying north over Wigford Down on 3 July.

1945: Two immatures were seen near Shipley Bridge, at Brent Moor House, on 17 April. They were also reported at an unspecified later date.

The records of 1934 to 1943 could have referred to the same returning individual, and it is interesting, if somewhat strange, that four of the records should have been in late spring or summer.

MARSH HARRIER *Circus aeruginosus*
Very rare passage migrant. Formerly bred.

It may come as a surprise to hear that this species, now so firmly linked as a breeding bird to the extensive reed beds and low land of the east coast, once inhabited many upland sites throughout Britain. As well as Dartmoor, Lakeland, Cardiganshire and Yorkshire were also inhabited (Holloway 1996). In Yorkshire they certainly bred on moorlands in the north and east up to the middle of the nineteenth century (Ratcliffe 1990). A national decline in breeding numbers occurred in the late nineteenth century.

The reason for this was relentless human persecution linked with probable loss of habitat through drainage.

D'Urban and Mathew report that the species was regular about rabbit warrens on Dartmoor up to the mid-nineteenth century and took a heavy toll of young rabbits. An account is given of a warrener who destroyed eight birds in one week during a severe winter. It would appear from this that birds were resident in Britain at that time, and not the mainly summer visitors we know today. However, a caveat must be given regarding the identification of these birds. Moormen distinguished this species from the Buzzard, but at this time of year could they have mistaken the species, and seen female Hen Harriers? By the late nineteenth century the nationwide decline was being noted on Dartmoor, and D'Urban and Mathew never encountered a Marsh Harrier in their many years of visiting the moor. It had by then become the very rare visitor we know today.

It has been suggested (Moore 1969) that considerable confusion occurred on Dartmoor in the nineteenth century over the identification of this species. This might well be so, but the decline of this bird as an upland breeding species in Britain co-incides exactly with the timing of its disappearance as a breeder from Dartmoor, and there appears to be no reason to doubt its status as given by D'Urban and Mathew and others.

Marsh Harriers fared badly in Britain for much of the twentieth century, and this was reflected in the lack of records for Dartmoor. It was only very recently with breeding numbers increasing and subsequent increase in the dispersal of post-breeding birds, that an upturn in Dartmoor records has been noted.

The total records are as follows:

1971: Prewley, 14 July, one flying southeast.

1975: Cadover Bridge, River Plym, 20 May, one female or immature.

1990: Okehampton, 3 February, an immature female flew north.

1993: Cadover Bridge, 14 September, one.

2001: Holne Moor, 11 May, one second calendar year female flew south.

2002: Buckfastleigh, 18 August, a female flew over.

2003: Crownhill Down, 3 August, an all dark juvenile flew south.

Swincombe Valley, 12 August, a female or immature.

Birch Tor, 2 September, a female or immature. Seen by same observer as 12 August bird, and thought to have probably been the same individual.

The early autumn of 2003 was exceptional for the number of Marsh Harriers passing through Devon and Cornwall.

An additional record came from the West Dart valley in the late 1990s, but the exact date was not noted.

HEN HARRIER *C. cyaneus*
Scarce winter visitor and passage migrant. Formerly bred.
According to D'Urban and Mathew the Hen Harrier formerly bred in the neighbourhood of Dartmoor. They thought that it still may have been resident, but was normally seen as a rather rare casual visitor, mainly in autumn. So by the late nineteenth century the Hen Harrier, probably never more than an occasional breeder on Dartmoor, had gone the way of the species in the majority of its former upland breeding areas. The decline, as with many other raptor species, was largely the result of human persecution.

The last breeding on Dartmoor, and in the county, quoting from D'Urban's statement in the Victoria County History of Devonshire (Moore1969), was at Torhill near Throwleigh, where a female and her four eggs were 'obtained' in about 1861. Indeed, as was usual at the time, most of the specific records referred to birds that had been shot. These include a male at Ashburton in June 1851, a pair at Chagford in 1871, and a young bird near Taw Head on 10 October 1877.

Throughout the eighteenth century it had been the opinion of most ornithologists that the grey Hen Harrier and the "Ringtail" Harrier were two different species. In 1805 Dartmoor played a small part in proving that to be untrue. In June of that year a nest with eggs was found near Bovey Tracey. When the young hatched Colonel Montagu obtained two and raised them into their second calendar year, noting that by this time one bird, a male, moulted from brown into basically grey plumage, and the other, a female, remained brown.

There have been no known breeding attempts since the nineteenth century, although at times birds have been noted during the summer months. There were summer sightings of birds in the 1940s, and there were rumours of a pair around the West Dart valley in the mid-1990s. The rumour was never proved but a definite sub-adult male was seen close by on the Swincombe on 23 June 1996. A ringtail was seen at Hameldown on 4 July 1993, and a male at Warren House Inn on 30 June 1999. What was almost certainly the same male was seen again near Two Bridges on 18 July 1999. With displaying pairs and summering birds on Exmoor recently (Ballance and Gibbs 2003), and a confirmed breeding record from Cornwall in 2002, there is the possibility that a pair may attempt to breed in the not too distant future.

Although a few early passage birds have been seen in August and September, it is October before migrants are usually found. Several of these are young birds in their first autumn. Dartmoor has a small but regular wintering population, and these birds are usually in the area by mid-November. Numbers then tend to stay fairly constant until March, although this can depend very much on the severity of the weather.

Birds will desert the high moor rapidly if there is snow, and in the last few years a reduction in the number of birds using a traditional communal roost has been noted after early January. This may well not be weather-related, and but only a case of birds moving to another, as yet unknown roost. Numbers drop rapidly after the end of March, but records are not too uncommon in April and even early May. These late passage migrants probably belong to one of the northern breeding populations, perhaps from Scandinavia.

In spring and autumn, birds hunting or on passage can be seen on occasions almost anywhere over the moor, but during the winter by far the majority of records come from the southeast area. With prey during the winter being in short supply on moorland, it is important for harriers to concentrate on productive areas. These areas are those with sufficient ground cover to offer food and shelter to the small number of ground-feeding passerines that still inhabit the moor at this time of year. Most of this productive habitat, with extensive heather and *molinia* covered bogland, lies in the southeastern portion of the moor, which also traditionally tends to be a little drier than the Atlantic-facing western side of the moor. Meadow Pipits feed and roost in these areas, whilst being very scarce on the open moor, and ground-roosting Fieldfares are regular in heather around Warren House Inn and the West Webburn Valley. Both are good food sources for hunting harriers.

The habit of communal roosting by Hen Harriers in winter had been remarked on in the literature since at least the early nineteenth century, but it was the descriptions of roosts in Dumfries and Galloway, excitingly described and illustrated by Donald Watson in *The Hen Harrier*, that really fired imaginations when it was published in 1977. By 1979 a winter roost had been located on Dartmoor, in deep heather in the vicinity of Warren House Inn. Although it has moved over the years, it is still active in the same general area today. The roost there has never been large. By 1980 up to five birds were known to be using the roost, and this remained the maximum number until possibly six birds roosted on 20 March 1994. This total was exceeded during January 2003, when up to nine or ten individuals were known to be visiting the roost. Unfortunately though, they did not all visit on the same nights.

Grey males have always predominated at this roost, with adult females being something of a rarity. The roost is used from October onwards, and it is probable that migrating birds use the roost ahead of the arrival of the wintering population. Wintering harriers, arriving from November, use the roost, but it has always been apparent that the same individuals do not roost every night. It is not fully understood why this is, but it is almost certainly connected with hunting patterns and areas used during the day. Weather is also a factor. More birds roost on mild, grey and windy nights than on cold, crisp, clear ones. Perhaps on fine, clear, days birds are hunting further afield and drop into roost at any likely site, rather than returning a long distance to the communal roost. Or perhaps on such nights they return late, after the light has gone, and too late for human eyes to note. The first birds arrive at the roost about half an hour before sunset, and the last ones usually about half an hour after

Hen Harrier

sunset. The few records we have of birds leaving the roost in the mornings, show them to be rather tardy, with some individuals only flying off several hours after dawn.

There have been one or two other roosting sites located on the moor, but the Warren House Inn area appears to be the only regular one. In the Scottish borders known individuals have been seen hunting up to 16 kilometres from their roosting site (Watson 1977). If this applied also to Dartmoor then it is possible that birds hunting almost anywhere on the moor could return on winter evenings to the heather by the Warren House to roost.

MONTAGU'S HARRIER *C. pygargus*
Very rare passage migrant. Formerly bred.

The Montagu's Harrier was first given species status in 1802 by the Devon ornithologist Colonel George Montagu of Kingsbridge, who proved without doubt that it was a species distinct from the similar Hen Harrier. He named it the Ash-coloured Falcon. Later in the century it was renamed by eminent colleagues in his honour.

At the time it was named by Montagu, it was not thought to be a bird breeding in Devon. Breeding was first proven in 1808. It never appeared to be plentiful, and throughout most of the nineteenth and twentieth centuries it was struggling to maintain its foothold in the county. By 1920 it was thought to be nearing extinction as a Devon breeding species (Smith 1956).

D'Urban and Mathew recorded breeding on Dartmoor near Ashburton in 1808 and 1809, both times in dense gorse brakes. They also mentioned that a pair were trapped at Bushworthy rabbit warren in October 1839. This location cannot be traced, and may be a misnomer for Ditsworthy or one of the other warrens in the Plym valley. A note from R.A. Julian published in *The Naturalist* in 1850 spoke of two being caught in gin traps on the moor (Jenks 2004). An adult male was killed at Trowlesworthy, another warren site, on 13 April 1872. On examination it was found that the bird had been feeding on Lizards, and the remains of fourteen were found in its crop. Lizards were also noted as prey at a Dartmoor breeding site in 1945 (Robinson 1949), although no doubt small passerines have always been more the norm. An adult male was shot at Christow in May 1870, and two melanistic birds were obtained at Lustleigh about this time.

In the early twentieth century pairs bred in certain upland bogs. The area beyond Archerton was mentioned for the 1930s and 40s (Dare and Hamilton 1968), and the area around Shoveldown/Long Stone/Kestor for a slightly earlier period (R. Waller *pers.com. 1997*). The latter area at that time had plentiful heather. However, the heyday of the species on the moor came when large areas of open moor were given over to forestry. Between the two World Wars and up until the 1950s many square miles around Fernworthy, Bellever and Soussons were planted with conifers. To enable the young trees to grow, areas were fenced off and made stock-proof. Dense areas of grass and gorse grew up between the young trees, and formed an impenetrable underlayer. This man-made habitat, quiet and undisturbed, proved very attractive to breeding Montagu's Harriers.

The three large areas of plantation all had breeding pairs from the 1940s until 1959, with one area having a short resumption of breeding in the mid-1960s. Fernworthy claimed the highest numbers with more than one pair breeding on occasions, and three in 1945 when pairs bred in close proximity. Two of the nests were only 20 yards apart (Robinson 1949). The best year was 1950 when four pairs bred successfully at plantation sites, producing seventeen young, thirteen of which fledged successfully. By the late 1950s breeding was not so regular and signs of a decline had set in. Birds arrived as usual in the spring, but increasingly did not stay to breed. The reason for this may have been habitat change. As the conifers increased in size, so the underlayer became more restricted and of less use to breeding harriers. There was a decline in numbers generally about this time in Britain and much of northwest Europe (Gibbons *et al* 1993), so other more widespread problems could have been affecting the small Dartmoor population. Chief amongst these must surely have been the insidious spread of pesticides through the food chain. The last successful pair bred in 1966 and two young fledged, but, ominously, this was also the year when a high level of Dieldrin was found in an egg.

Nearly all the young from Dartmoor nests were ringed as nestlings. Recoveries of birds were as follows:

1949: Nestling female ringed on Dartmoor in 1948, found on Anglesey 18

August with damaged wing.

1950: One ringed on Dartmoor found dead under wires at Meymac, Corrèze, central France on 8 September.

1955: A female picked up dead near Frome in the third week of September had been ringed on Dartmoor as a young bird on 17 July 1953.

1958: A bird ringed as a nestling on Dartmoor in August recovered near Brucheville Manche, France on 7 September.

After the passing of the breeding population, the status of the species on Dartmoor declined rapidly to one of irregular passage migrant, and in recent decades it has declined still further to be a very rare migrant. Since the last breeding in 1966 there have only been 25 records from the moor, and nine of these were before 1970.

In 2000 a national influx of birds occurred in mid-May. It was probably caused by continental birds over-shooting the Channel in adverse weather conditions. This influx provided Dartmoor with three records, an unprecedented yearly total in recent times. An adult male hunted the Plym valley around Spanish Lake and Blackabrook on 5 May, and an adult female was seen over Roborough Down on 12 May. On the 18 May a second calendar year male was found near Warren House Inn, not far from the old breeding localities. It stayed in the area until the 27 May, and interestingly used the same area of deep heather to roost as the wintering Hen Harriers.

A second calendar year female hunted the Warren House Inn area from 3 until at least 9 June 2002, and the latest Dartmoor record was of a ringtail watched hunting at Black Ridge on 25 June 2004.

GOSHAWK *Accipiter gentilis*
Very scarce resident breeder, and passage migrant.
The Goshawk was admitted into the Fauna of Devon on the authority of Dr. Tucker of Ashburton who claimed it had been found on Dartmoor (Bray 1836). It must always have been a very scarce bird, as few further early records of any validity exist. It was, however, thought to have bred just outside the area at South Tawton in 1830. D'Urban and Mathew quoted the above records, but had very little faith in them. They thought the occurrence of the bird in Devon was very doubtful, and square bracketed the species.

It was not until the mid 1980s that this species became established on Dartmoor following re-introduction in Britain in the 1960s and 1970s. The first successful breeding was not proven until 1989. Since then a small population of a few pairs has built up where suitable habitat exists. Site details are not published to give this rare bird a chance to breed in peace, as they are very vulnerable to disturbance and prone to persecution. Despite known pairs showing good productivity (mean 2.9 young per pair 1991-2003), further expansion appears very slow. Perhaps one reason is the fairly recent felling of large areas of the upland forests of Dartmoor, both limiting preferred habitat and causing disturbance. Indeed at the time of writing two of the original occupied areas now appear unutilised.

Although not normally breeding until at least their third calendar year, one of the earlier breeding attempts in 1994 involved a second calendar year female paired with an adult male. The pair was successful, fledging two young, and the event was repeated at another site in 2003.

Analysis of prey at Dartmoor sites shows a diet comprising mainly Carrion Crow, Jay, Magpie, Woodpigeon, Grey Squirrel and Rabbit, indicating that perhaps Goshawks do more good than some may think. No game bird remains have ever been found at a Dartmoor nest site (M. Darlaston and R. Khan *pers. comm* 2004).

SPARROWHAWK *A. nisus*
Resident breeder and possible passage migrant.
Sparrowhawks breed regularly and in good numbers in broad leafed woodland around the periphery of the moor, as well as the conifer plantations of the interior.

Although they made no specific references to Dartmoor localities, D'Urban and Mathew remarked that it had formerly been the most abundant species of hawk in Devon, but at their time of writing (1892) although tolerably numerous, it was much persecuted. As the twentieth century progressed, Sparrowhawks appeared to be doing quite well away from keepered estates, where they always had a hard time. In 1951 H.G. Hurrell knew of at least six nests within a three mile radius of his home on the southern edge of Dartmoor (Moore 1969).

Then came the national devastation of DDT and other pesticides. Dartmoor, and Devon generally, was spared the almost total wipe-out of the species suffered in counties on the east coast, but its numbers were still greatly reduced, with poor breeding success from the remaining bird population. With the banning of many of the offending pesticides, Sparrowhawks quite quickly showed that they were regaining ground. At this time, the mid-to late 1960s, birds would also have been able to take advantage of some of the plantations approaching maturity and providing more breeding habitat.

Breeding was suspected at Bellever in 1957 and 1958. By 1966 two pairs were proved breeding with another pair at Runnage in 1965 (Dare and Hamilton 1968). Thirty years later an estimated five pairs were breeding in plantations at Bellever and Soussons, with an additional pair in the lower West Webburn Valley around Cator (Dare 1996). Direct comparisons like this are not available for other areas, but every indication shows that the species is doing well. Three pairs were displaying over woods by Trendlebere Down on 11 April 1999, and again 20 February 2000, and two pairs were over Soussons Plantation on 7 March 1999.

On a BTO Common Bird Census plot of 76.7 hectares at Harford, comprising mostly of moor edge farmland, a single pair was present from 1990 to at least 2000. The Dartmoor Study Group has noted an increase in records in recent years, with birds being noted at a maximum of thirty sites in 2000.

Birds will occur over open moor, well away from their woodland haunts, on occasions. Sparrowhawks are opportunist hunters, and will utilise open habitat when the need arises, especially during the breeding season. A bird hunting low south of

Okement Hill, in the middle of the north moor, on 1 May 1997, was probably taking advantage of the abundance of young Skylarks and Meadow Pipits. A female actively hunting over the open moor at Down Ridge and Skir Gut on 15 February 2001 could well have been drawn there by the newly arrived Skylarks taking up territories.

The status of the species as a passage migrant is difficult to assess with conviction. Birds passing over high on fine days in August and September could be migrants from northern Europe, but also, and perhaps more likely, they could be juveniles dispersing from not too distant natal areas.

BUZZARD *Buteo buteo*
Common resident breeder, and scarce passage migrant.

To many people the sight of a Buzzard soaring over a valley on a fine day epitomises Dartmoor. D'Urban and Mathew made little comment about them as far as Dartmoor was concerned, but did note that they used to see a few on the outskirts of the moor. They listed two traditional nesting sites near Fingle Bridge and at Gidleigh Park, with the inference that, by mentioning individual sites, they were by no means common as breeding birds. They mention that Buzzards were more plentiful as migrants in spring and autumn, but perhaps they were misreading the pre- and post-breeding parties seen at these times of year.

The numbers we see today were certainly unknown in the nineteenth century. As with so many raptors, this was probably due to persistent persecution from game keepers and farmers. By the early 1900s Buzzards were increasing, helped no doubt by the lack of keepering in the First World War. In *Victoria County History of Devonshire* (1906) D'Urban mentions seven pairs in the Teign Valley above Steps Bridge in 1911, a precursor to the general increase some years later. The population continued to do well until 1954 when myxomatosis decimated the rabbit population of Britain. Having lost their staple prey Buzzard's breeding numbers declined rapidly, and in the worst areas a third of the population came to grief (Moore 1969).

The birds breeding on and around Dartmoor escaped the worst food shortages. In the Postbridge area there was a small reduction following the outbreak, but numbers were stable again by 1960 with 10/15 pairs. There was a temporary reduction again in 1963, when only eight pairs bred following high mortality the preceding winter (Dare and Hamilton 1968). Sixteen or seventeen pairs were breeding in the area by the early 1990s (Dare 1996). In line with the Postbridge findings, the general population on Dartmoor stabilised quite quickly after myxomatosis, and increased from the early 1960s. Some noteworthy records involved seven nests around 6.5 miles of northern Dartmoor near Okehampton in 1976, four breeding pairs in a wood of five acres at Hennock in 1985, and two pairs in a BTO Common Bird Census site of 76.7 hectares at Harford, 1987 to 1994, with three pairs 1995 to 1996, four in 1997 and five in 2000. It would be difficult today to think of a wooded Dartmoor coombe or plantation that does not have breeding Buzzards. Few estimates of the total Dartmoor population have been made, but in the 2000 survey for the Dartmoor National Park the numbers were thought to be stable, and were estimated at 351 pairs +/- 76.

Birds are usually seen soaring singly or in pairs, but in early spring pairs are re-establishing territorial boundaries, and some quite large gatherings are seen as neighbours gather to display over their territories. Twenty over Moretonhampstead on 12 March 1995, and seventeen over Trendlebere Down on 27 March 1999, were good examples of this. Post-breeding parties can also occur, where parents from adjoining terrirories soar with their offspring. Examples of this were the 15 over Western Beacon on 29 September 1972, the 14 over Legis Tor on 22 September 1992, and the 19 near Ashburton on 9 September 1995.

Migration has not been noted with any certainty in the spring, but definite autumn passage over or around Dartmoor has been observed on several occasions. Two instances are particularly worthy of note. In 1980, at a time of unusual Buzzard movement in southern Devon, at least 27 birds were watched on the 24 September at Dousland, coming off the moor at a great height and drifting away southwest into the wind (Dare 1999). On the 23 September 1992 seven Buzzards were noted soaring together near Cadover Bridge on the River Plym. They all soared to about 100-150 metres then drifted off east southeast. Later other small groups appeared from the west and as they reached the northwest slopes of Trowlesworthy Tors they too soared up and drifted away east southeast. The total amounted to at least thirty seven, and the observer had the clear impression that they were all moving in a definite direction and following the same course (M. Williams *pers. comm.* 1993). Interestingly, although not noted elsewhere in Devon at that time, there had been an influx of birds into West Cornwall on 20 September, with as many as 95 birds recorded, including 35 at Land's End. The Dartmoor birds may have been a part of this influx of migrants, moving around the southern edge of Dartmoor, and then on eastwards to locate a more acceptable Channel crossing.

Another, possibly unique gathering that may have included migrants occurred in September and October of 1991. On 13 September 29 birds were counted in the area of Harford Moor and Ugborough Beacon. They increased almost daily and by 20 September 44 birds were in the area. There was then a gradual decrease until the last 10 were seen on 15 October. The reason for this remarkable congregation was thought to be the huge flight of Crane Flies (*Tipulidae*) that was occurring at that time, and proved to be a useful food source for several weeks.

The plumage of the Buzzard is notoriously varied, but it would be impossible to leave the subject of Dartmoor Buzzards without mentioning one particular bird. This was the female that became famous as the White Buzzard of Buckfastleigh. This pure white bird arrived at Biggadon Farm, Buckfastleigh in February 1957. It had been reared some four miles away at Westcombe Farm, Dartington in 1952, and had stayed there until 1956, eventually moving, perhaps in the aftermath of myxomatosis. It stayed in its new-found home all the rest of its life, becoming tame enough to accept the presence of the farmer, and actually allowing him to ride underneath the tree in which it was perched. It became something of a celebrity, and finally died of debility

Common buzzard

due to old age on 19 September 1973, being twenty one years old (Hurrell 1974).
Not the only white Buzzard to appear on Dartmoor, it was certainly the best known,
and probably the longest lived.

ROUGH-LEGGED BUZZARD *B. lagopus*
Very rare winter visitor and passage migrant.
D'Urban and Mathew knew this species as a rare winter visitor that only occurred in
years when there had been an influx of Scandinavian birds onto the east coast. They
mention the winter of 1875 as being especially good for the numbers that arrived in
Devon, the 1876/77 winter was likewise excellent. A bird was shot, and another seen
at Ditsworthy Warren on 31 January 1877, and a female was trapped near Lustleigh
on 28 March 1883. Two more were said to have been obtained on Dartmoor in the
mid-nineteenth century, but no further details were given.

It must be assumed that as four of the above records referred to birds 'obtained',

61

then the identification based on feathered tarsi was correct. But based purely on sight, many of the nineteenth and early twentieth century records would not be found acceptable today. The differences in plumage, structure and size compared with Buzzard were not then appreciated. The following are the more recent records accepted by the Devon Bird Watching and Preservation Society, of what is now a very rare visitor to Dartmoor.

> 1970: Dendles, one migrant stayed for several days in late March.
> 1978: Dartmeet, 22 October, one.
>   Princetown, 3 December, one.
> 1979: Soussons, 30 December, one.
> 1995: Amicombe Hill, 3 May, one. This migrant appeared after a winter that produced a large influx, mainly on the east coast.

OSPREY *Pandion haliaetus*
Rare passage migrant.

The first recorded Osprey on Dartmoor would appear to be the bird shot on the river Meavy at Goodameavy in September 1905. For many years this individual, mounted in a very dramatic pose, was on display in Plymouth Museum, and is no doubt remembered by many of the city's birdwatchers. There were then no further records until 1952 when an immature was found at Fernworthy on 11 September.

Between then and 2002 there have been sightings of thirty birds. The earliest was one at Hennock on 8 March 1979, and the latest an individual that stayed in the Venford Reservoir and River Dart area until 8 October 1993. Most spring records have been in mid-April to mid-May, and the autumn records from the third week of August to the end of September.

As would be expected, most birds were seen at reservoir sites. Fernworthy and Burrator have six records each, and Hennock three. Most birds did not stay for more than a single day. The two exceptions to this were the bird that stayed at Fernworthy from 20 September to 3 October 1987, and the Venford Reservoir and River Dart bird that stayed from 16 September to 8 October 1993. All records have been of singles except two watched circling and moving north over Grenofen on 7 August 1998.

There would appear to be a tendency for birds to fly around the moor, rather than over it. Several of these moor edge records have been linked with river valleys.

In addition to the Meavy record mentioned above, there have been another three in the general area. Around the Buckfastleigh and Hembury areas of the Dart, birds have been noted on four occasions. Migrants have twice been watched skirting the edge of the southern moor at Ugborough Beacon, and a bird was seen going west around the edge of the high northern moor at Belstone. In the more central areas of the moor, a bird was watched fishing the Cherry Brook at Powder Mills on 29 May 1965, and another was seen over Princetown flying towards Burrator on 24 September 1978. The only record over the high northern blanket bog was of a bird flying southeast at Okement Hill on 3 June 1997.

KESTREL *Falco tinnunculus*
Resident, and possible passage migrant.

D'Urban and Mathew made no specific comments pertaining to Dartmoor, but thought the species a generally distributed resident in Devon that was tolerably common. Its fortunes have taken a down turn in recent years, and although the Kestrel could still be described as generally distributed as far as Dartmoor is concerned, its numbers could no longer be described as tolerably common, as many areas appear to have lost breeding birds.

Kestrels increased nationally in the early twentieth century, probably as a result of decreased persecution (Holloway 1996). This increase was probably reflected in the Dartmoor population, but records for this period are lacking. In the Postbridge area, an increase was noted from the early 1960s. One or two pairs, probably breeding in 1956-58, increased to two pairs of confirmed breeders most years from 1963, with a further one or two pairs nearby in the West Webburn Valley, also from 1963. The population then fluctuated between one and five pairs, but with a probable peak of seven pairs in 1978 (Dare and Hamilton 1968 : Dare 1996).

The last county avifauna, published in 1969, stated that the Kestrel was Devon's most numerous raptor by far (Moore 1969), but by the time of the *Tetrad Atlas*, a decline was being confirmed by the fieldwork, carried out between 1977 and 1985. Both Buzzard and Sparrowhawk were found to be more numerous in the county (Sitters 1988). As would be expected, most Dartmoor registrations were for peripheral areas, with birds seeming to be more plentiful on Exmoor. This decline was noted nationally through the BTO *Common Bird Census* after 1981.

Although not well documented, this decline continued into the 1990s. At Postbridge the species was noted as amazingly scarce in 1990, and still very scarce in 1991, with only two or three singles seen and breeding very doubtful. 1992, although slightly better, produced only one pair in the area, despite six scattered birds being seen in March. Elsewhere around Dartmoor, comments on the scarcity of birds at this time were noted in *Devon Bird Reports* for Yarner Wood, Buckland Common, Pudsham Down, Hameldown and North Bovey. In 2000 the *Dartmoor National Park Authority Survey* estimated through extrapolation that the population for the moor was 38 +/- 10 pairs. At this time the average number of sites recorded by members of the Dartmoor Study Group per year was 57. The average for breeding season records/sites was 32, thus broadly in line with the findings of the 2000 *Survey*.

It is not as yet proven that the Kestrel is a passage migrant on Dartmoor, although it is likely. Young birds disperse from July onwards in their first calendar year, and it is known that in the breeding population of upland areas of northern Britain, adults and young move away in the winter, with a return movement peaking in March and April (Wernham *et al* 2002). Some of the birds noted in the Postbridge area in March 1992 could have been migrants. Unfortunately, the difficulty of being able to distinguish between migrants and wandering local birds is so great that the answer may never be forthcoming. It does look probable that Dartmoor residents do move down to lower altitudes in extreme winter weather. The records of the Dartmoor Study Group consistently show fewer birds in the winter months, especially January and February, and an altitudinal movement has been suggested in these circumstances at Postbridge (Dare 1996).

At times in late summer and autumn family parties can be seen hunting together over the open moor. Normally these parties consist of five or six birds, but occasionally numbers present would suggest that two or three families have come together to take advantage of a particular food source. The nine at Moorhaven on 8 October 1989, the ten at Black Tor Copse on 12 August 1972, and certainly the record of twenty one hovering over a ridge at Walkhampton on 3 August 1977 would suggest this.

RED-FOOTED FALCON *F. vespertinus*
Very rare vagrant.
Only two records of this falcon.
A female was watched at Soussons on 20 June 1970, and a second calendar year male was present at Burrator from 18 to 20 May 1999. On the last date it was watched flying off strongly to the east up the Newlycombe Lake Valley, and was not seen again.

MERLIN *F. columbarius*
Scarce passage migrant and winter visitor, and very scarce former breeder.
On the 16 May 1920 F. Howard Lancum flushed a female Merlin from a ground nest amongst heather and other herbage near Cadover Bridge (Lancum 1934).

It contained four eggs and was the first confirmed breeding record for Dartmoor. The observer noted at least one other breeding attempt in the area in subsequent years, and believed that the species was less rare than was generally supposed. He had seen Merlins in gamekeeper's 'galleries' in woods on the southern edge of Dartmoor at least a dozen times, and in June 1933 had found a pair nailed up in a plantation some three miles west of Cadover Bridge, which could have been on the eastern edge of Roborough Down. The next known breeding was in 1934 when W. Walmesley White confirmed a pair using a deserted Buzzards nest in a willow tree on the South Teign River (Walmesley White *Ornithological Journal* (Isabelline, 2004). The location must have been near or just inside the present day Fernworthy plantations, just north of White Ridge. So Dartmoor had birds nesting in tree and ground locations at an early date.

D'Urban and Mathew mention that Merlins were said to have bred in Hound Tor Vale, and that young had been taken from a nest on Roborough Down, but did not give much credence to either report. Interestingly, the Roborough Down reference may not have been too far from where Howard Lancum found the shot pair forty years or so later.

It is more than probable that Merlins bred on the moor prior to the earliest confirmed record, and continued to do so after the second 1934 breeding, but further definite records are lacking for this early period. There were several records in the 1940s and early 50s of birds in the summer months in an area of the Plym valley around Ringmoor Down, Gutter Tor and Ditsworthy Warren. Although nothing was ever proved, breeding in the area was more than possible. The area is only a short distance from the site near Cadover Bridge where the pair nested in 1920.

The next sequence of confirmed breeding took place from 1953 in the plantation areas of Bellever and Fernworthy. At Bellever, and presumably at Fernworthy, pairs bred in old Crow and Magpie nests (Dare and Hamilton 1968). One or two pairs bred most years at Bellever up until 1959, with occasional records into the mid 1960s. At Fernworthy pairs were also successful up until 1966. After this date birds have been seen on many occasions during the breeding season, at several localities, but breeding has never again been proven.

Today Merlins are seen on passage and in the winter, when individuals can occur regularly in some areas. They range widely and have large hunting territories out of the breeding season, with birds being seen wherever there is sufficient cover for their prey, mainly Meadow Pipits, to feed. The area around Warren House Inn, Vitifer, Birch Tor and Merripit Hill holds birds most winters. Walkhampton Common and Leedon Tor is also a regular spot, as is the area around Haytor and Rippon Tor, and the Swincombe and West Dart areas.

The first autumn passage birds usually appear in late August or early September, quite often associated with Meadow Pipit migration. The bulk of the passage birds arrive in late September and October, and by November birds that remain usually winter on the moor. It would be interesting to know the number of wintering birds, but their wide ranging hunting and large territories make assessment of this very difficult. It must certainly run into double figures for an average winter. Birds stay until March, and some into April. Late April and even early May records no doubt refer to northward bound migrants.

As well as their usual hunting techniques of direct aerial pursuit and surprise, birds have on occasions been seen hunting in conjunction with Hen Harriers. The Merlin uses the harrier to flush prey and then attempts to rush in and secure it before the more ponderous harrier can react. On at least three occasions Merlins have been seen to hover, an activity that is not well documented. There have been three interesting records concerning the recovery of ringed birds. In 1957 a bird ringed as a nestling on Dartmoor was found dead at Copplestone near Crediton during the summer, obviously not long after fledging. Another from the same brood was recovered on 9 September at Serigny, Andilly (Charente-Maritime) France. On the 29 June 1990 a dead Merlin was found at Sharp Tor. It was ringed, and proved to be the first Icelandic-bred bird found in Britain in summer. It had been ringed in northeast Iceland as a nestling on 4 July 1984 (John 1990).

Hobby eating dragonfly

HOBBY *F. subbuteo*
Scarce summer resident breeder, and passage migrant.
In their introductory paragraph on this species D'Urban and Mathew stated that it bred occasionally in the woods on the borders of Dartmoor. They then listed sites at Gidleigh Park, where birds had bred for several years about 1870, Woodtown near Horrabridge where a pair bred in 1871, and Lydford Woods. The earliest Dartmoor breeding record that they had noted was a female killed on her nest at Brimpts on 12 July 1849.

Hobbies still breed in the same type of border woodland, but in recent times they have been able to take advantage of the maturing conifer plantations. In the Postbridge area it was formerly a rare summer wanderer, but has been increasingly reported since 1970 (Dare 1996). The same situation exists in many other localities. Hobbies can be extremely elusive in the breeding season, but there is no doubt that they have increased greatly in the last thirty years or so. Breeding sites are kept secret for security reasons, but it is probable that the number of pairs breeding in the Dartmoor area runs into low double figures.

Birds arrive from mid-April, with most in late April and May. Some may be passing migrants rather than breeding birds. Especially early birds were those at Yelverton on 7 April 1999, and Hennock Reservoirs on 9 April 1982. Once both

66

birds have arrived, and a nest chosen (usually an old Crow's nest) breeding starts in earnest, and after initial display they usually go very quiet. The best chance then of seeing a Hobby is on any warm afternoon or evening in late spring or early summer, when birds can be watched hawking for insects over areas of heather moorland.

After breeding, with the young fledged, family parties will stay in the area for some weeks before migrating south during September. At this time prior to departure they can appear quie noisy, with the young calling loudly. Most birds leave in early to mid-September, with some, perhaps juveniles on their first migration, staying until later in the month. Occasionally birds can be seen into early October, although these could well be migrants from further north, and not local breeders. Two late birds were the individuals in the Glaze Valley on 1 October 1991, and at Whiddon Down on 2 October 1996.

GYR FALCON *F. rusticolus*
Very rare vagrant.
There is only one record. In the Fernworthy area, a dark phase bird was present from 18 to 20 February 1985. During its stay it was watched plucking a Teal that it had killed.

PEREGRINE *F. peregrinus*
Scarce resident breeder. Passage migrant and winter visitor.
During the late nineteenth century, although Peregrines were met with not infrequently throughout the county, they were very much birds of the coast, and were rarely found far inland. So stated D'Urban and Mathew. Although always subject to drastic persecution, numbers increased in Devon in the twentieth century until the first major setback to hit the species occurred during the Second World War. As they were thought to constitute a threat to carrier pigeons, many were culled by the Air Ministry. No sooner did numbers start to pick up after the war than another setback hit the country's birds. The decimation caused by toxic chemicals has been fully described elsewhere, but suffice it to say that Devon's Peregrines declined almost to the point of extinction. When the use of the offending chemicals was banned or at least greatly reduced, numbers again began to increase. By the late 1970s the recovery was well under way, but few at that time would have thought that coastal sites would ever prove insufficient for the expanding population. By the mid-1980s the first Devon inland sites were being tenanted, and in due course the Peregrine became a breeding bird on Dartmoor.

Because of the risk of disturbance and persecution no details of breeding sites are given. However it can be said that the 2002 *Peregrine Survey* found twelve territories either inside or overlapping the National Park boundary. With suitable coastal territories fully utilised, and inland sites now in very short supply, will Dartmoor Peregrines eventually be forced to use tree or ground sites?

With the number of breeding pairs in the area, wandering unattached individuals, and dispersing juveniles, Peregrines can be seen over the moor just about anywhere, and at any time of the year.

RED GROUSE *Lagopus lagopus*
Resident breeder.

The Red Grouse is a species introduced to Dartmoor. D'Urban, updating the county list in 1906 for the *Victoria County History of Devon* (1906), square-bracketed Red Grouse as not worthy of confirmed admission. Little is said about them in the very earliest *Devon Bird Reports*, and probably the best account of their early days on the moor is given in Douglas Gordon's *Dartmoor in all its Moods* (1931). In this, Gordon explains that details of the first introduction are lost in the mists of time. They were of course introduced for shooting, and the earliest scheme dates to the early nineteenth century. Nothing is known regarding the outcome of this introduction, other than that it failed. The scheme that Gordon had personal knowledge of happened in 1912, when one hundred pairs were released onto the moor. By 1920 they were thought to be firmly established, but no organised shoots had taken place because it was thought prudent to let the breeding population increase further first. Also in 1920 a further one hundred pairs were released to boost the breeding stock. However, from about this time a decline in fortunes was noticed. Gordon suggests that the cause of this was the reduction of good heather habitat, caused by an increase in the number of Scottish black-faced sheep on the moor, coupled with excessive swaling. By the early 1930s big shoots still had not taken place, and the Duchy of Cornwall, under whose auspices the introduction had taken place, had largely lost interest in the idea. Birds were now becoming scarce, and to illustrate this Gordon describes a walk from Belstone to Fur Tor and Dart Head in the autumn of 1930,where he saw or heard no bird until flushing a pair with a late brood at Cut Hill. Poaching and foxes were also given as reasons for the lack of Red Grouse at that time. But the main cause was without doubt the lack of sufficient good heather areas where the birds could thrive and spread.

The 1933/34 *Devon Bird Report* contained a note from Gordon confirming a decrease on north Dartmoor, although he thought them more abundant south of the river Teign. The picture he painted for their future, was however, a gloomy one. Broods were small, family party sizes were consequently greatly reduced, and he thought that there was no hope of them 'making good' under the conditions then prevalent. This was another reference to the woeful state of the heather.

Between that date and the present time the fortunes of Red Grouse on Dartmoor have always been linked with the abundance and condition of heather, the birds main nesting habitat, shelter in the winter, and food source. Although certain areas prove more acceptable some years than others, and there are natural peaks and troughs in the overall population, the numbers during the past seventy years or so appear to be fairly stable. *Devon Bird Reports* over the last forty years have shown an average of 24.36 occupied localities per year. The maximum number of localities was 37 in 1969, and the minimum 15 in 1981. There is of course nothing to suggest the level of observer coverage and this may have varied widely from year to year. One observer noted birds in 22 localities in 1988.

Professional surveys of the moor give a somewhat truer picture, but are not much at odds with figures calculated from the *Devon Bird Reports*. In 1979 a complete moor survey was carried out for the RSPB. This survey recorded a total of 57 'pairs', which amounted to 24 confirmed or probable breeding, and 33 possible breeding. The importance of the heather was again noted and found to be present in 23 of the 24 confirmed or probable territories, and giving an average ground cover of 45%. Purple Moor Grass *molinia* was another important constituent of the habitat, being present in 19 of the 24 territories, and giving an average ground cover of 22%. Most territories were on damp rock-free plateau or gently sloping areas of blanket bog or fringing heather moor (Mudge et al 1981). In 1992 a follow up survey for the RSPB covered a portion of the original 1979 survey area. This area, which had held 17 territories in 1979, and 19 in an additional survey in 1986, held 23 in 1992. The area consisted of 52 square miles of high moorland on the north moor (Chown et al 1992). A High Moorland Sample Survey for the Dartmoor National Park Authority in 1997 found four territories in an area of 16 square kilometers. This compared with ten in the same area in the 1979 Survey, five in 1986 and 12 in 1992. The latest survey for the Dartmoor National Park Authority in 2000 found seven males in a 5 square kilometer area, which through extrapolation gave an estimate of 18-30 territories for the moor (Geary et al 2000).

Over the years *Devon Bird Reports* have noted birds in some 80 localities on the northern moor. As in Gordon's day by far the best area has always been the high blanket bog habitat of Okement Hill, Cranmere Pool, Hangingstone Hill and Whitehorse Hill. Other localities with many records have been High Willhays, Cut Hill, and Cosdon Beacon. On the eastern side of the moor, another area that has consistently provided records, albeit at a much lower level, has been the heather moorland around Warren House Inn, Vitifer, Birch Tor and Hameldown. Most sightings have been in the Hameldown area. Between 1956 and 1967 up to five pairs were known to have bred in this eastern area (Dare and Hamilton 1968), and more may have done so before the severe winter of 1946/47. In the southern part of the moor records are far more scattered, presumably because of the poor denuded heather in the area. Birds have been recorded at some 50 localities on the southern moor over the years, but numbers have always been low. The areas with the most records have been Ryders Hill, Eylesbarrow, and Nuns Cross.

Normally the most a walker will see of Red Grouse is one or two birds whirring away and calling loudly. Occasionally, at the right time of year, a family party may be seen, but now and again, perhaps after a good breeding season, a larger party will be encountered. The largest parties on Dartmoor have been 14 at Postbridge on 21 October 1948, 11 at Cut Hill on 24 September 1975, Cosdon Beacon on 28 October 1982, and Hameldown 19 November 1983, and 10 at Ockerton Court on 12 November 1998.

BLACK GROUSE *Tetrao tetrix*
Formerly a scarce resident breeder. Now extinct.

The Black Grouse was almost certainly indigenous to Dartmoor, although introductions, none of them well documented, have clouded the picture somewhat.

From as early as the late eighteenth century writers had remarked on the diminishing numbers of Black Grouse on the moor. Moore, writing in 1837, suggested that they could soon become extinct, and Rowe a few years later stated they were fast disappearing. At the end of the nineteenth century D'Urban and Mathew, although still stressing the decline and scarcity of the species, did not talk of imminent extinction. Perhaps captive-bred stock had been released on the moor in the intervening years since Moore's comments.

The reasons for this decline, which also affected all the other isolated populations in southern England, are not fully understood. On Dartmoor it was thought that poaching was largely to blame. Miners and turf cutters were known to take birds, and there are records of moormen running down Blackgame with dogs. This activity proved that, given the right circumstances, they were not too difficult to catch. In view of modern knowledge, it would appear far more likely that some environmental factor was the major cause. Over-stocking with cattle and sheep and the consequential loss of heather, the bird's main food source, could well have been an important reason. Excessive swaling also perhaps played its part in the decline. By the 1930s, with the population still only just holding on despite introduction and re-introduction schemes, a short-lived new lease of life was given to Black Grouse by the introduction of conifer plantations onto the moor. In their early stages these areas, first around Fernworthy Reservoir, and later in the 1940s, around Bellever, provided excellent habitat for the remnants of population. However as soon as canopy was formed these areas became unsuitable. The breeding population, no doubt by then already irretrievably doomed because of its low numbers, became extinct very quickly. The last male at Bellever was seen in 1954, and although there were records of birds, especially males, for a few years after that, the centuries of breeding Black Grouse on the moor had ended.

The major areas where birds were located in the first half of the twentieth century, were as follows:

Fernworthy and the north moor: Although birds were occasionally seen over much of the high northern moor during this period, it was only the conifer planting around Fernworthy Reservoir that gave a reprieve to the small breeding population. A 'lek' was established here in the 1920s on Assycombe Hill (R.Waller *pers.comm.* 1997), and although never very large it persisted in the same general area into the early 1950s. The maximum number of males displaying at these gatherings was seven on 25 April 1942. The last bird was recorded there on 1 August 1953. It is probable that an additional, or alternative, 'lek' site was used at nearby Merripit Hill in the early 1940s (Dare and Hamilton 1968).

Bellever: A blackcock was first recorded in the Bellever area on 1 June 1937, and it, or an additional male, was in the Laughter Tor area on 6 June. In the late 1940s a stock of about 25 birds was released in the area (Dare and Hamilton 1968). By 1948 a 'lek' was being recorded in fields adjacent to the young plantation, to the west of the Youth Hostel (L. Slade *pers.comm.*1997). At its height this 'lek' held up to seven displaying males, with the maximum being seen on 24 April 1949. This population was never prolific, and numbers fell rapidly in the early 1950s. The last recorded male was seen on 13 May 1954. Apparently, the last birds were shot by forestry staff.

The Plym Valley and the south moor: During the 1940s and 50s there were a considerable number of records from areas on the southern moor. The origin of these birds must be open to question. They may have been survivors of the old wild population or wandering birds from Bellever or Fernworthy. Alternatively, they could have been part of some unrecorded introduction scheme.

The nucleus of this population appeared always to be in the Plym valley, especially around the source area, and the area to the north including Nun's Cross, Hand Hill, and Fox Tor. The first record would appear to be six seen in the Plym valley on 9 May 1917 (the late H.G. Hurrell *pers.comm.* 1980). This early date, well before known introductions at Fernworthy and Bellever, lends some weight to the argument that numbers could have existed here from the earlier indigenous population. This was also the largest number ever encountered on the southern moor. Birds were also found regularly around the head waters of the Erme and the Yealm, occasionally lower down the Erme valley, and to the east as far as the Avon. Usually only one to three birds were seen, but at times slightly higher numbers were recorded, as with the five at Royal Hill on 8 March 1950. Numbers tailed off as the 1950s progressed, and probably the last record was the male flushed by the present writer near Yealm Head in the January of 1958.

Since the demise of the Fernworthy and Bellever birds in 1953/54, and the southern moor birds a little later, records became very sparse. Ironically, the biggest single party seen on Dartmoor, away from the 'leks', occurred at this time, when nine were watched at High Willhays on 9 November 1957. During the late 1950s and 1960s birds, usually singles or pairs, were reported from Beardown, between Amicombe Hill and Little Kneeset, Okement Hill, Gidleigh Moor and Yes Tor.

After this time the very few records tend only to be accepted with the caveat that the origins of the birds were very much in doubt, and could well have referred to stock captive-bred and released onto the moor. Between 1970 and the present birds have been seen at High Willhays, Lydford Forest, Vitifer, Cosdon Beacon, Rough Tor, Huntingdon Warren and Kitty Tor, where the most recent bird was seen on 26 April 1997.

QUAIL *Coturnix coturnix*
Very scarce summer visitor to Devon, that very rarely occurs on Dartmoor.
This migrant has always arrived in the county in very variable numbers, but 'Quail Years' have in recent times been very few, and we no longer experience years, as in the nineteenth century, when birds could be widespread and quite abundant.

Dartmoor could never be considered as good habitat for this species, which favours dry rough grassland and large fields of arable crops. In almost all cases, birds have been recorded on Dartmoor only in years when numbers in the county generally have been well above average.

Most of the few records have been from agricultural land, with the area around Bridford being the only locality to have birds in more than one year. Calling birds were located there in June and July 1976, June 1994, May 1997 and June 2001. In 1987 a bird was heard at North Bovey on 31 May, and one was killed by a cat in Okehampton on the late date of 3 November. A little closer to the moor proper, two were reported from a moorland area near Tavistock in 1966, and one was heard calling at Fernworthy on 10 June 1976. The most unusual record, however, is that of a bird found in the centre of the northern moor at Great Kneeset on 22 June 1992.

As elsewhere with this species, birds are usually located by their calls, and are rarely seen. Breeding is thus extremely difficult to prove, and given the unsuitable nature of most of Dartmoor it must be very doubtful if Quail have successfuly bred on the moor.

PHEASANT *Phasianus colchicus*
Resident around moor edge, where large numbers are artificially reared and released.

Up until quite recently it would have been true to say that Pheasants, raised in increasing numbers on moor-edge estates, rarely venture on to the moor.

In the Postbridge area small numbers released in the lower West Webburn Valley over the years have all failed to establish a population, possibly due to predators. The bird's status there is no more than a scarce resident that possibly breeds occasionally (Dare 1996). In the West Dart valley around Huccaby, stock is put down each year in March, and sightings then increase appropriately through the spring and early summer, although breeding probably only occurs in very small numbers (Hibbert 2000, Hibbert 2001). In another area where birds have been monitored for many years at Harford, a Common Bird Census site of 76.7 hectares has shown great variability in the number of territories from year to year. These no doubt reflect the number of birds put down and the extent of shooting each year, but even then the variance is extreme. The 19 territories in 1989 was an increase of 170% over the figure for 1988. Territories were counted as 23 in 1990, and 33 in 1991. In 1992 the figure had dropped to just 16, and over the next seven years the number never exceeded 15, and indeed in 1998 only reached seven.

The above three instances show the general difficulties Pheasants have with the area. They do not generally move far from their area of release, and appear quite happy to exist as shotgun fodder. Even allowing for the large numbers lost to guns each year, it does seem surprising that some small populations have not built up quietly in likely areas.

In the last few years, perhaps because of an increase in releases on certain estates and the subsequent pressure on feeding areas, birds have appeared further into the moor. Birds are now quite regular in many areas around Burrator, including the Deancombe valley, where they used to be very rare. Others have been seen by the Blackabrook, upstream from Cadover Bridge, around Fernworthy Reservoir, and in Gutter Tor Marsh. In 2000 several were flushed on Penn Moor, up to 500 metres into the open moor. All represent relatively short distances from traditional habitat, but perhaps in time we may find this species even further from its 'in country' home.

RED-LEGGED PARTRIDGE *Alectoris rufa*
Scarce visitor, from stock reared for shooting on moor-edge estates.
This has always been a species introduced to Devon. D'Urban and Mathew mention several records, but there are no convincing ones for Dartmoor. The first record for the moor would appear to be the six pairs that were released on the Delamore estate at Cornwood in September 1979. About the same time, birds were released in the area of Roborough Down, from the Maristow Estate.

Many of the Dartmoor records since then have been close to these two estates, with birds being noted quite regularly at Cornwood, Harford, Lee Moor, Portworthy and Shaugh Prior. At the Harford Common Bird Census site pairs were noted breeding in 1992 and 1999. In other areas birds were recorded at Haytor Quarries and Postbridge in 1981, Fernworthy in 1993, Gidleigh Common in 1998, Leusdon in 2000, and Drewsteignton and Okehampton in 2001. In recent years the numbers being released appear to have increased considerably, with a consequential increase of birds at moor edge sites. In 2001 birds were reported in six areas. About 20 birds seen in fields adjoining Roborough Down in December showed that there was plenty of potential for birds to wander into other areas. However, even the edges of Dartmoor are not ideal habitat. Any pairs that do breed successfully in the wild are likely to be few, and so birds are likely to always be infrequent away from their artificial rearing areas.

GREY PARTRIDGE *Perdix perdix*
Formerly a wild resident breeder, but now a rather rare visitor from released stock.
D'Urban and Mathew noted them as increasing in Devon in the late nineteenth century, and only being absent from the wilder parts of Dartmoor.

Being one of the stalwart species of the countryside they were long taken for granted, and thus very few notes appear as to when their decline in numbers began. But a decline there certainly was, and it gained pace in the last few decades of the twentieth century until the species disappeared as a wild bird from many areas. This was generally thought to be linked with the change of farming practice after the Second World War, when a far more intensive system was adopted.

On Dartmoor the decrease has been very largely unrecorded. The only information comes from the Postbridge area (Dare and Hamilton 1968; Dare 1996) where the species was noted up to 1968 as a sporadic visitor to the lower farmed land. Two pairs appeared in the late summer of 1960 at Broadaford in the West Webburn valley. They stayed throughout 1961 but breeding was not proved. One pair bred in 1962, and birds were noted in the area up until 1965. In 1961 a pair bred successfully at Postbridge, but were not seen again. However, a pair noted by the East Dart River above Bellever on 14 June 1962 could have been the same birds. None were recorded after 1968.

In recent years the few that have occurred have no doubt come from released stock, with the last of the wild birds disappearing quietly many decades before. Birds have been seen between 1989 and 2001 at Ashburton, Dunsford, and Roborough Down. One particularly interesting record, because of its distance into the moor, was a single at Hexworthy on 2 July 1998.

GOLDEN PHEASANT *Chrysolophus pictus*
Rare escapee from captivity.
A bird adopted the garden of Torgate House at Princetown during March and April 1996. It was strange that it had chosen a high and rather inhospitable area of the moor for its temporary home. Another male was seen in Sticklepath, usually on the cricket field, between April and June 2002, and the same year a male was killed by a car on Shaugh Moor.

WATER RAIL *Rallus aquaticus*
Very rare winter visitor and passage migrant.
D'Urban and Mathew commented that the Water Rail was said to breed in the neighbourhood of Dartmoor, although they had never seen one on the moor. In view of the great rarity of birds on Dartmoor in the twentieth century, this comment on its former breeding status must be open to doubt.

Extremely elusive, it is probable that wintering individuals could be missed in marshes and along moor-edge streams. In any case the known records are very few. The six confirmed records are as follows:

1951: Green Hill, 25 December, one almost in the centre of the southern moor. A most interesting record, and one that poses the question of how many others have been missed over the years in similar habitat.
1980: Manaton, 7 November, one found dying in a house.
Burrator, 11 November, one.
1983: Princetown, 10 December. one.
1986: Powder Mills, 19 September, one on the Cherry Brook.
1999: Sampford Spiney, in January the remains of one, recently dead, found in a marshy area.

In view of the fact that the British breeding population is thought to be resident, the few Dartmoor records are probably of continental migrants moving south and west to winter.

SPOTTED CRAKE *Porzana porzana*
Very rare winter visitor and passage migrant.
D'Urban and Mathew had never flushed any on Dartmoor, and confidently stated that the country was too high and exposed. They did mention one at Bridestowe on 1 October 1862, but this may have been just off the moor. There is also a reference to one at Postbridge in 1862 (Hurrell 1947). Other than these the only record was of a bird flushed beside the West Okement near Black Tor on 12 to 15 December 1969.

LITTLE CRAKE *P. parva*
Very rare vagrant.
A female of this species was shot near Ashburton in 1809 and sent to Colonel Montagu for identification. For many years it was considered to be the first for Britain, but an earlier record came to light of a bird obtained at Catsfield in East Sussex in 1791, thus making the Ashburton individual the second for Britain. Until 1890 the Sussex bird had mistakenly been labelled a Spotted Crake (Palmer 2000).

CORNCRAKE *Crex crex*
Very rare passage migrant. Formerly bred in small numbers, and was a regular, and at times reasonably common autumn passage migrant.
No bird epitomises the decline of species in Britain due to changes in agriculture, better than the Corncrake. A century or more ago they were regular breeders in hay-meadows and the like throughout most of Britain, including Devon.

On Dartmoor, their breeding areas were more the agricultural 'in country' than the moor itself, although the open moor was frequently inhabited by migrants, especially in the autumn. D'Urban and Mathew commented that on Dartmoor they found birds in the autumn on the most elevated bogs, such as Raybarrow Mire, and sometimes encountered six or more in a morning's walk when they were Snipe shooting. August 1877 was noted by them as a particularly good period, with birds numerous on the wet bogs.

The decline was first noted in the late nineteenth century. Even at this early stage writers linked the decrease with the mechanisation of mowing. The mechanical mowing was indiscriminate, destroying many nests and young, when previous methods might have spared them. Being quicker, mechanised mowing also meant that bigger areas were covered earlier in the season, thus putting more pairs and nests at risk (Holloway 1996). The decline was noted to take hold in the south and east of the country first.

By the time that detailed records were published in Devon in the late 1920s and early 1930s, the county population was already in a terminal decline. Records for Dartmoor, be it prospective breeders calling in spring or even migrants in autumn, were getting progressively rarer year by year. Of the few records at this time, one was of a corpse found amongst bracken at the head of Deancombe valley, Burrator, in August 1921 (Smaldon 1982). It had been freshly killed by a Sparrowhawk. Two possible breeders were heard calling at Mary Tavy in 1933, but only one was heard in 1935. During the 1940s birds were still being found, but usually they were noted with a comment as to their scarcity. A bird calling at Buckfast on 16 April 1947, was the first heard in the area for ten years. A bird was driven from cornfields at Uppacott, east of Chagford, on 16 and 17 July 1948, suggesting that breeding may have taken place, and a bird was calling at Manaton on 25 May 1949. The same year a migrant was flushed from long grass near Hexworthy on 23 August.

The 1950s produced the same sprinkling of records, but with one possible exception, they were all migrants. One was disturbed on Holne Moor on 3 September 1951, another was heard calling near Ashburton on 13 May 1953, and a further bird was flushed the same year from long grass at Postbridge on 18 September.

One called at dusk in pastures at Runnage on 24 April 1958, but was considered to be a migrant (Dare and Hamilton 1968). Another migrant ran across the road at Haytor on 19 September 1959. There were only three records during the 1960s but as they were all of calling birds in spring it is possible that at least one may have constituted a breeding attempt. In early June 1964 a bird was heard calling on several occasions near Tavistock, and another was heard calling in the same area on 17 April 1966. The third record was of one at Chagford on 10 May 1969.

From 1970 until the present day birds have been very rarely encountered, and all have been autumn migrants. A bird was caught and killed by a terrier on the edge of Royal Hill on 13 September 1976, and the remains passed to the present writer for identification. A migrant was at Wrangaton on 22 September 1979, and another at Holne on 25 September the same year. The latest Dartmoor record would appear to be the bird rescued from a cat at Scorriton near Holne on 5 October 1990.

MOORHEN *Gallinula chloropus*
Rare resident breeder.

The small population of Moorhens on Dartmoor are heavily reliant on farm ponds and the like for breeding sites. The fast rivers of the moor are not to their liking, and most of the reservoirs are not found acceptable, perhaps due to their acidity and lack of sufficient suitable food.

In the 1960s it was noted as a vagrant in the West Webburn valley with only two records (Dare and Hamilton 1968), but during the 1990s there was an attempt at colonisation, with pairs appearing at several ponds and breeding proved in 1993, 1997 and 1998. A private Nature Reserve pond at Throwleigh had a nesting pair in 1990, and each year 1997 to 2000. Other ponds have held pairs in recent years at Chagford, Horsham Pond, Moortown (Sampford Spiney), Aller Farm (North Bovey) and at Sourton Cross where three pairs bred in 2001.

Of the reservoirs, Fernworthy has the best record, with pairs breeding in 1987 to 1990, and probably 1998 and 1999. Hennock Reservoirs had three breeding pairs in 1993 and one in 1996. Birds are seen occasionally at Burrator, but breeding has never been proved. An unusual occurrence took place in 1996, when a pair nested and raised five young on the River Meavy, just below Meavy village.

At Huccaby, where the species is noted as a winter visitor only, one stayed until 22 March 2000, and was seen again from 17 September, and on several dates through to 6 April 2001. One, perhaps the same bird, was back at the end of August 2001, and was still around at the end of the year. Two were seen on 2 October 2001 (Hibbert 2000 and 2001). A juvenile seen on the Becka Brook near Hound Tor on 17 September 2001 was an example of a bird wandering away from its natal area.

COOT *Fulica atra*
Winter visitor and passage migrant. Normally quite rare. Has bred.
Although almost exclusively a winter visitor, birds are seen at times in spring and autumn, and on one occasion at least birds have remained to breed. A pair nested at Hennock Reservoirs in 1996, and a breeding attempt may have been made there in 2000.

Hennock is also the only reservoir that attracts wintering birds on a regular basis, irrespective of the severity of the weather. Ten to twenty have been usual here each winter since the early 1960s. Maximum counts at Hennock have been 35 on 17 February 1962, 40 on 17 January 1970 and 45 on 13 February 1974. However, a startling 250 were counted in severe weather on 9 January 1945. This was by far the largest count for Dartmoor. Numbers at Burrator are normally very low, but numbers can rise due to cold weather influxes into the area. The maximum here has been 40 in mid-January 1982, at a time of snow and bitter cold.

Other reservoirs normally attract on ones and twos, although a count of 17 in March 1998 at Meldon was unusual, and probably represented a flock on passage forced down by inclement weather, as this is not a regular wintering site. Occasionally birds will turn up on quite small areas of water, as with the bird that stayed on a private nature reserve pool at Throwleigh on 26 and 27 March 1996, and the single found on the small reservoir at Swincombe Intake Works on 23 November 2000.

LITTLE BUSTARD *Tetrax tetrax*
Very rare vagrant.
D'Urban and Mathew listed a bird shot at Ashburton in about 1820. This would appear to be the only record for the Dartmoor area, although there is a mention of a bird seen in the Plym valley by the Dartmoor historian R.H. Worth (Hurrell 1947). No further details or date was given.

GREAT BUSTARD *Otis tarda*
Very rare vagrant.
Once again, the inclusion of this species on the Dartmoor list rests on just one record. D'Urban and Mathew quote Colonel Montagu as saying that a bird was killed near Plymouth in the winter of 1798, and two more the following year. One of these two was shot at Houndale, Dartmoor in 1799. Houndale has proved somewhat difficult to trace, but may very well refer to Houndle or Houndall, on the southern edge of the moor, west of Cornwood.

OYSTERCATCHER *Haematopus ostralegus*
Rare passage migrant.

There have been about seventeen records of Oystercatchers on or over Dartmoor since 1950. Of these, eleven have occurred between mid-July and the end of August. They perhaps represent juveniles moving away from their natal areas, possibly on the Welsh coast. They could well have been heading for the Exe Estuary to join the nationally important numbers that winter there.

Nine of the records at this time come from the Okehampton/Meldon area, suggesting a small regular passage either over the moor, or perhaps around its westerly slopes. Most records have been of single birds, some seen flying south, but two were at Meldon Reservoir on 15 July 1984, and a party of six put in an appearance there on 20 August the same year. Other areas have had records in this autumn period, but only Fernworthy Reservoir has had birds in more than one year. Singles were there on 17 September 2000 and 14 July 2003. Elsewhere, at least three were heard overhead flying south-southeast at Wrangaton at 10.35 pm on 19 August 1956, and a single bird was disturbed from the North Teign river, opposite Quintens Man on 18 August 1987. Slightly later in the autumn, calling was heard at Wrangaton on the dark foggy night of 3 October 1950.

There have been two winter records of birds passing over the moor. One was recorded flying south at Prewley, near Okehampton, on 7 December 1977, and three birds were observed flying south off the moor at Ugborough Beacon at dusk on 22 November 1992. The remaining two records could well have referred to migrants leaving wintering grounds and returning north to breeding areas. A bird was seen at Fernworthy Reservoir on 20 February 1988, and calls were heard over Wrangaton on the night of 14 March 1963.

One record is something of a mystery. On 1 June 1996 the remains of an Oystercatcher was found at Winneys Down on the northern moor. It appeared to be a fairly recent Peregrine kill. The bird had been ringed as a nestling on Skomer Island, Dyfed on 18 June 1993. It thus tended to confirm the idea that Welsh birds could migrate over Dartmoor, but the date was late for return spring passage, and as this bird was reaching maturity it could have been expected to be closer to its natal area, where most birds return to breed. What was it doing on Dartmoor at this time?

BLACK-WINGED STILT *Himantopus himantopus*
Very rare vagrant.

In what must surely rate as the most bizarre of Dartmoor bird records, one was found feeding along a stream near Buckland Beacon on 25 March 1990. It was subsequently seen near Haytor, where it stayed from 26 March to 2 April.

March 1990 was remarkable for the number of southern European species that arrived along the south coast. Among these birds were Black-winged Stilts, that arrived in unprecedented numbers and at amazingly early dates. At least 20 were involved in the influx, which began on the 17 March and produced records from southwest

England, south Wales and southern Ireland.

D'Urban and Mathew mentioned that this species was on the list of Dr. A.G.C. Tucker as having been seen in the Ashburton area in the early nineteenth century. No details are given, and it is impossible to say whether the bird was within the Dartmoor area.

AVOCET *Recurvirostra avosetta*
Very rare vagrant.
Although Avocets have wintered in ever increasing numbers on the Tamar estuary since the late 1940s, there have only been three records for Dartmoor. Two records are from Burrator and would appear to be birds travelling to and from winter quarters on the Tamar. A single was seen there on 14 November 1965, and another on 26 February 1986. The third record was of a bird at Portworthy Dam on 17 February 2002. This, too, was likely to have been an individual dispersing from wintering grounds on the Tamar.

STONE CURLEW *Burhinus oedicnemus*
Very rare vagrant.
D'Urban and Mathew mention two birds shot on Dartmoor in October 1831. One was killed at an undisclosed location on 5 October, and the other at Widecombe a few days earlier.

In more recent years the only record is of a bird seen well at rest and in flight at Cut Hill on 5 September 1958.

LITTLE RINGED PLOVER *Charadrius dubius*
Rare passage migrant, and recent breeder.
With only two records of birds at Burrator on 3 May 1969 and 24 April 1984, it looked likely that this species would always maintain the status of a rare passage migrant.

Then, in 1998, birds began to be seen on the industrial china clay pits of the Lee Moor area. Two were present in March of that year and a single bird between 1 and 7 July. The next year birds again appeared on 24 April, 11 June, 11 and 19 July and 31 August. In 2000, two birds, perhaps a pair, stayed from 3 to 9 April, and an adult and a juvenile were noted on 17 July. Either the same or another two birds were there on 23 July. It seemed just possible that given the similarities between the occupied habitat and their man-made breeding sites over much of east and middle England, a pair may just have stayed to breed. 2001 however, proved a little less encouraging, with singles only being seen on 2 May, 19 August and 2 and 16 September.

Things looked decidedly better in the spring of 2002. Four birds, thought to be two pairs, were present in late May. Display was noted, and a nesting attempt looked a real possibility. Disappointingly, things went quiet in June, and it was thought that although one pair was still present, any eggs laid had been claimed by predators.

However, in mid-July the remaining pair was watched with four young, obviously a successful replacement clutch. The adults and three or four young were seen on a number of occasions into August. This constituted the first breeding record for Dartmoor and Devon.

Birds appeared again in the spring of 2003, but did not appear to breed. Time will tell whether the exciting events of 2002 were a one-off or the first stage in colonisation of the region.

RINGED PLOVER *C. hiaticula*
Rare passage migrant.

Very few records have been noted for this species. The earliest, in spring, was also the most unusual of the Dartmoor records. It involved a group of three sheltering from the weather on open moorland near Hangershell Rock, Harford on 31 March 1999. Another interesting spring record was of a bird flying down the West Okement at Okehampton on 21 April 1982. The only other spring record is of three at Portworthy Dam on 19 May 1999.

Birds have been noted about twelve times on autumn passage in August and September. The area of Lee Moor clay pits has proved a favoured habitat, with birds in several years 1993 to 2001. Usually one or two were involved, but occasionally small parties occur such as the seven present on 22 August 1999 and the five there on 19 August 2001. Fernworthy is the only reservoir to have records in more than two years. Six were there on 22 and 23 September 1996, with three still present on the 26. This coincided with an exceptional passage of small waders on the moor, and through Devon generally at this time. Ones or twos were also noted at Fernworthy in September 1999 and 2001.

The only other records consist of one at Burrator on 27 August 1955, with two there on 15 October the same year, and a single watched at Meldon Reservoir on 5 September 1995.

DOTTEREL *C. morinellus*
Rare passage migrant.

Due to the confiding nature of this species, their ability to merge with the landscape, and their unwillingness to take flight, they can very easily be overlooked. This, coupled with the fact that their lengths of stay, in the spring especially, are usually very short, could mean that Dotterel are under-recorded and perhaps pass through Dartmoor in small numbers every year.

D'Urban and Mathew knew them as very rare stragglers to Devon, but were able to quote two records of multiple numbers. Several were seen on Chittaford Down, near Postbridge, in September 1828, and many were shot on Dartmoor during April 1840. They also remarked that although they had received specimens from North Devon, more had been obtained in the south of the county, around the southern slopes of Dartmoor. In recent times these southern slopes, grass covered and Ploverfriendly, have still proved attractive to Dotterel.

The most frequented site on the moor has been the ridge between Yes Tor and High Willhays, with sightings in the spring of 1990 and 1999, and the autumns of 1969, 1998 and 2001. Sites having records in two years have been Cox Tor, Okement Hill, and Hameldown. Areas on the northern moor that have had records in one year only have been Hangingstone Hill, Sourton Tors, East Okement Farm, Okehampton Common, Devil's Tor and Cosdon Beacon. On the southern moor birds have been seen at Harford Moor, Dean Moor, Three Barrows, Hawks Tor, Western Beacon, Ugborough Beacon, Royal Hill, Greencoombe, Widecombe and Roborough Down.

The earliest spring record was of two birds on High Willhays on 7 April 1999. Other early records in the spring were singles at Three Barrows on 11 April 1988, and Greencoombe on 15 April 1989. These early birds appear to have been associated with migrant Golden Plover flocks moving north. The last ten days of April, and the first week of May, is the optimum time for spring Dotterel passage over the moor. It is not so much that a large number of single birds are seen at this time, but this is the period when 'trips' have appeared. The first was a totally unprecedented flock of 22 that was seen on Royal Hill on 21 April 1984. The second was a 'trip' of six birds seen together between Yes Tor and High Willhays on 30 April 1990, and the third record was of six on 21 April 1997 in the area of East Okement Farm. 2004 turned out to be an exceptional year, with a 'trip' of five at Devil's Tor on the 5 May, and six at Widecombe about the same time. There have been a further four records of singles up until the middle of May, and one record of a late migrant on 25 May 1975. Very unusual was the record of a bird, possibly a female, on Roborough Down on 10 June 1992.

In autumn the earliest recorded was a bird on Western Beacon on 24 August 1996. Passage birds have occurred on eight dates up until mid-September, and then there have been three records up to the end of the month. Birds, usually juveniles, tend to stay a little longer in autumn, as with the three on High Willhays 27 August 1998, that stayed until 1 September, and the three on Cosdon Beacon on 10 September that stayed until the 13 September. There have been no October records, but a very late bird was seen with the Golden Plover flock on Cox Tor on 10 November 2001.

GOLDEN PLOVER *Pluvialis apricaria*
Very rare and declining migrant breeder, passage migrant and winter visitor.
The Rev. Thomas Johnes writing to Mrs Bray in the early 1830s, stated that locals had told him that Golden Plover had bred in Fox Tor Mire (Bray 1836). This area has had a few records in the breeding season in more recent years, including one bird present in May 2004, although breeding has never been proven.

Although D'Urban and Mathew quote several comments regarding breeding Golden Plover from other observers, they were themselves unconvinced, and thought they did not breed any further south than the Breconshire mountains.

There were no records of even possible breeding in the early twentieth century.

Douglas Gordon, writing in 1931, considered it only a winter visitor to Dartmoor, and it was not until 1948 that a pair was found in suitable nesting habitat on the northern moor (Nelder 1949). Birds were again seen in the summer of 1949, but it was 1950 before breeding was proved. On 19 June that year Dr. L.H. Hurrell watched a pair with three young at an undisclosed location on the northern moor. During the next few years the northern blanket bog received more attention, and pairs were found on several occasions, although proof of breeding was rarely confirmed. Locations were sometimes not made public, but Cranmere Pool, Hangingstone Hill, Cut Hill, Okement Hill, Taw Head and West Dart Head all figured in sightings during the 1950s and early 1960s.

By the mid-1960s careful searching was turning up an increased number, and up to four pairs were not unusual. In June 1967 one observer found at least six pairs in the Walkham Head area alone. Details of records in the *Devon Bird Reports* at this time are not very helpful. It is impossible to hazard a guess as to how many pairs actually held territories, as locations were only given in a very sketchy way, if at all. With only a very few interested people involved in the fieldwork, no full survey of likely areas was ever undertaken. The same minimal information was given in the 1970s, but as some of the published records concerned multiple pairs in rather restricted areas, e.g. three pairs on Black Ridge on 16 June 1973, it is probable that the totals for the moor at least went into double figures. In 1979 the RSPB and the Dartmoor National Park Authority jointly initiated a full survey of breeding bird populations on the open moorland of Dartmoor. This survey found 14 Golden Plover territories in the area from Cranmere Pool to Cowsic Head. Thus the first real benchmark for measuring the fortunes of species on the moor was obtained. Since then their fortunes have been all down hill. Follow-up surveys in 1986 and 1992 showed only nine and six territories respectively.

In 1989 John St. Leger commenced his regular yearly hunts for breeding pairs. Together with Mike Sampson, he provided nearly all the information up until the *Upland Bird Survey* by the DNPA in 1999. By the early 1990s the old areas around Cranmere Pool, Black Hill, Okement Hill, etc. were being deserted, and the small number of pairs present were mainly to the south, in the area around Cut Hill, Cowsic Head and Tavy Head. Between three and five territories were located most years in the 1990s, and although 1994 proved a little better, with six to eight, the general decline continued. Even the professional survey of 1999 could only find five territories. Only one or two pairs were present between 2002 and 2004, and it looks inevitable that the Golden Plover must soon become extinct as a breeding bird on Dartmoor.

The causes of this decline may be many and varied. The 1992 *Survey* put forward the view that an increase in the height of vegetation, due to lack of grazing, could have made large areas of formerly acceptable habitat unsuitable. Human pressure in the last quarter of the twentieth century must also take a fair proportion of the blame. However, Golden Plover have always been on the extreme southern edge of their range on Dartmoor, and must always have been vulnerable.

All confirmed and probable breeding has occurred in the blanket bog areas of the

Golden Plover

northern moor. There have been sightings of birds during the summer months on the southern moor, but records of even probable breeding are lacking. Occasionally during their breeding season, off-duty birds fly outside their territories to feed. They then turn up on short grassland areas, like Ringmoor Down and particularly the area around Lynch Tor. Birds frequently behave quite territorially in these areas if others are present, and this can lead to incorrect conclusions regarding birds breeding away from the blanket bog. It is probable that most records on the southern moor refer to wandering off-duty birds, non-breeders or late spring passage migrants.

Breeding birds and their young stay around the moor into August, and it is probable that the records of small flocks at this time are made up of local birds. In the 1950s and 1960s the fields around Powder Mills were a regular place for these early flocks. Migrants from further north begin arriving in September, with the large flocks usually arriving in the last week, and into October. Small flocks can then be found almost anywhere over the moor where suitable areas for feeding exist, but there are certain traditional areas that hold larger flocks. The grass moor around Cox and Staple Tors regularly holds the largest flock on Dartmoor. Records here go back to at least 1958. The flock size varies, but usually peaks between 1500 and 2000 birds. Small flocks break away from this number, and can be found feeding around Princetown and in the Peak Hill/Sharpitor area. Another large flock can normally be found in the Harford Moor/Ugborough Beacon area. This too has been known to reach 2000, although in recent years 500/800 has been normal. Other regular sites for flocks numbering in hundreds exist around Haytor/Rippon Tor, Ringmoor Down, and the Okehampton area.

If the winters are mild, the flocks will stay throughout. This in itself is at variance with most other upland areas in Britain, where birds depart for lower ground at the onset of winter. Once hard frosts or snow hit the high moor, flocks depart at once, but will return quickly after a thaw. Departure of wintering birds usually begins in March. On 26 March 1994 two flocks totalling 800 to 1000 birds were watched flying northeast off the moor over Cosdon Beacon.

A passage of migrant Golden Plover going north takes place during April. Many of these migrant flocks consist of birds showing breeding plumage of the Northern race (*P. a. altifrons*). The passage of northern birds can continue well into April, with the last flock of any size being 200 on Harford Moor on the very late date of 3 May 1970.

GREY PLOVER *P. squatarola*
Very rare winter visitor and passage migrant.
Only two records exist of this most coastal of plovers. In very severe weather conditions in 1986 one was sheltering at Burrator on the 19 February. An autumn passage bird was seen feeding on Ringmoor Down on 29 September 1994.

LAPWING *Vanellus vanellus*
Scarce and declining breeder, passage migrant and winter visitor.
For a species that is so much part of the British countryside there is an unfortunate scarcity of information on the population of earlier years. On Dartmoor, as in most of Britain, they were just too common to be noteworthy. D'Urban and Mathew, writing at the end of the nineteenth century, thought they were increasing annually, but made little other comment.

Early *Devon Bird Reports* contain very few details of breeding numbers, and it is not until the Postbridge survey years (Dare and Hamilton 1968) that any conclusions regarding numbers of breeding birds are reached, albeit on only a small portion of the moor. The number of pairs in the Postbridge area fluctuated between 15 and 25 in the 1956-62 period. After the severe 1962/63 winter, none bred the following spring. One pair was back breeding in 1964, and then five to ten pairs by 1967. Ten to 12 pairs were breeding in the early 1990s (Dare 1996).

In 1979 the RSPB/DNPA survey of the whole moor found a total of 54 pairs (Mudge et al 1981), but this total was thought to be a little on the low side. Breeding was confined to the lower areas of the moor with none found over 1600 ft. Follow up surveys by the RSPB of part of the northern moor in 1986 and 1992 found no pairs and seven pairs respectively. This was in an area where seven were found in 1979. The monitoring of sites was undertaken more regularly in the 1990s, and the overall impression was one of continued decline. The total number of pairs located on the moor during this time was between eight and 18, with a diminishing number of sites occupied. Some of the deserted sites, like the marsh on the western side of Wigford

Down, had been occupied since at least the mid-1940s (L.W. Slade *pers. comm.* 1997). As all sites had never been covered in the same year, the Dartmoor Study Group carried out a full survey of known sites in 1999. The result was a total of 16 to 19 pairs, with most on the eastern side of the moor, particularly the area of Blackslade Mire and Bonehill Down, with others at Peter Tavy, Thornworthy, Cator Common, and Wigford Down. A Moorland Breeding Bird Survey, carried out for the DNPA in 2000, found 11 to 16 pairs. No birds were found in 28 of the 32 one kilometre squares, where at least one pair was present in 1979. The decline since 1979 was 65-70%, and was most noticeable on the western side of the moor.

The decline, in this once common species, has been nationwide. The full reasons are as yet still unknown, but in much of the country loss of suitable habitat has been blamed. This would not appear to hold true for Dartmoor. Unchanged habitat is still present in some abundance, so other factors must be involved. Increased visitor numbers in recent decades has no doubt had a detrimental effect, either directly by disturbance, or indirectly by forcing birds to leave the nest and thus making predation by Crows, etc., that much easier. It could well be, as mentioned in the 2000 *Survey Report* (Geary 2000), that even in 1979 the population was already below a self-sustaining level. Poor breeding success may also have been implicated in the decline. More research is needed to fully understand the failing fortunes of the Lapwing.

British-bred Lapwing are partial migrants, and after fledging birds will flock locally for a time before moving away. Some Dartmoor birds may not move far, but others may join the numbers wintering in France and Spain. One interesting record concerned a bird ringed as a chick near Merrivale in June 1980, and recovered in northern Spain in January 1987. In the 1960s and 1970s late summer flocks were frequently located in areas on the west side of the moor. In 1969 there was an early flock of 300 at Prewley on 8 July, with the same number there on 17 July 1970. A flock of 450 was close by at Meldon Reservoir on 28 July 1976. These flocks no doubt consisted of local breeders, with some immigrants from further afield. With declining numbers these flocks now appear to be a thing of the past.

Autumn always used to bring further migrants onto the moor, and even during winters, if the weather was not severe, flocks could be found at favoured spots, for example 500 at Prewley on 3 October 1971, and 400 at Lee Moor on 29 January 1984. In more recent years numbers have dropped considerably. An example of this is the regular small wintering flock at the Prison Fields, Princetown. Up until the early 1990s this regularly peaked at about 100/150, but recently the figure has been down to 40 or so.

Lapwing leave wintering areas at the first onset of heavy frosts or snow. On Dartmoor, as elsewhere, these conditions can lead to spectacular weather movements. About 1000 flew east at Sampford Spiney in a blizzard on 21 February 1978. Severe weather in January and February 1979 in the north and east brought many hundreds into the area with many large counts around the southern edges of the moor, the largest being 700 at Wrangaton on 20 February.

KNOT *Calidris canutus*
Very rare passage migrant and winter visitor.
There are just two records to date on Dartmoor, and both from Portworthy Dam. In 1999 a bird was present on 21 December, making it the first inland winter record for Devon, and the first record at any time of year for Dartmoor. Another bird was seen on 21 September 2002. This bird, an immature, stayed until the next day.

SANDERLING *C. alba*
Very rare passage migrant.
Portworthy Dam scored again with this species, with one in summer plumage on 16 May 1998, and an incredible flock of 17 that dropped in during heavy rain on 26 May 2000. There is often a mid-to late May passage of this high Arctic breeding wader, but these records are exceptional.

LITTLE STINT *C. minuta*
Rare passage migrant.
The September of 1996 was exceptional for this species over much of Britain. In Devon birds turned up in numbers at just about every location imaginable, including some on Dartmoor where they had never been seen before.

The first was a bird at Cadover Bridge clay pools on the 8 September. A single was seen at Portworthy Dam on 18, followed by seven on 21. Numbers at Portworthy had dropped to two by the 23 September, but one remained until 13 October. At Fernworthy Reservoir two were present on 22 September, but the number had increased to eight on 23. Burrator Reservoir, where any wader is a rarity, unexpectedly had the highest number for a moorland site with 11 on 25 September, although these quickly passed on as none were there on 27.

Since 1996 the only records have been from Portworthy Dam, where one was present on 10 September 1998. This number increased to three by 14 September.

PECTORAL SANDPIPER *C. melanotos*
Very rare vagrant.
The only record for Dartmoor of this transatlantic vagrant was the bird present at Portworthy Dam from 30 August to 3 September 1999.

CURLEW SANDPIPER *C. ferruginea*
Very rare passage migrant.
1996 was an especially good year for this species in the county, as it was with Little Stint. Birds were widespread at suitable localities, but unlike Little Stint only a few reached Dartmoor. The favoured site was Portworthy Dam, where three were present on 8 September, and two on 13 September. These were the first recorded on the moor. The only other sighting occurred in 2003, when two immatures were present at the disused clay pit pool at Leftlake on 6 August.

DUNLIN *C. alpina*

Very rare migrant breeder, and passage migrant.

The story of Dunlin on Dartmoor almost totally concerns breeding birds, as although passage migrants do occur, they are decidedly scarce for such a numerous species.

D'Urban and Mathew mention them as probably breeding in small numbers on Dartmoor, and record that they saw a breeding-plumaged bird in June 1866. They also quote Dr. E. Moore as stating in 1837 that a nest had been found. There is little information from the early twentieth century, but there is an intriguing list of sightings almost every year from 1937 to the mid-1950s, most of which certainly refer to breeding birds. However, it was not until 1956 that breeding was at last proved.

Most of the early records were in the area of Cranmere Pool, Taw Head and East Dart Head, although there was a record from Rough Tor in June 1955, and another from Conies Down Tor in early August 1947. The confirmed breeding in 1956 involved a nest with three eggs found at Taw Head on 24 June. Two eggs were still in the nest on 9 July, but the outcome was not known. Birds were again reported in the area in 1958, and in 1960 two pairs were found on 20 June, and a nest with two chicks and one chipping egg was located. Pairs, and at times nests, were found most years in the northern blanket bog until the early 1980s, when dwindling observer coverage made the status of birds uncertain. The best years appeared to be 1969, when about ten males were seen and heard on 1 June, and one nest located. 1967, with seven or more pairs, and 1970 when at least seven territories were occupied, including one with a nest and young in a new area. It is probable that the totals of the good years resulted from increased observer coverage, rather than from increase in breeding numbers.

In 1989 John St.Leger commenced regular visits to the northern moor to search for Dunlin. Because of his hard work and that of Mike Sampson, knowledge of breeding Dunlin numbers returned to earlier levels. Three probable, and one possible territory was found in 1989, with five probable and four possible territories in 1990. Most were in the traditional areas of Cranmere, Black Hill, Okement Hill and West Dart Head, but other pairs were being found further south in areas around Cut Hill and Cowsic Head. From then on there was a tendency for searchers to find more territories away from the old northern area around Cranmere. Indeed, in some years none were found there. Recently the southern area around Cut Hill, Cowsic Head and Tavy Head has turned up more pairs, and new sites like Wildbanks Hill have come into the picture. Just as if to prove how fickle breeding birds can be at times, on 10 June 2000 Mike Sampson found an adult with three chicks on Okement Hill. This was in the area where territories had declined in recent years. It was further north than the observer had ever seen birds, and was only a mere 200 metres from the heather boundary. During the last ten years numbers have varied from a low of six in 1995 to a high of 15 in 1999. This could be due to the ground and weather conditions at the time, but with a species that is far from easy to locate, it could also show certain variations in observer coverage from year to year.

Certain professional surveys have taken place over the years. The results have largely confirmed the findings of independent observers. Twelve pairs were located in 1979, 14 in 1986, and 10 in 1992. In 1999, Bob Jones, surveying for the DNPA, located a total of 15 territories, the highest total ever (Jones 1999). The conclusion must be that unlike Golden Plover, Dartmoor's population of Dunlin is holding its own, although always susceptible to changing conditions and disturbance. It is the most southerly breeding population in the world, and Dartmoor would be much the poorer without these interesting birds of the high blanket bog.

All confirmed breeding has been on the northern moor, but there have been a few interesting reports from the southern moor over the years. A pair in breeding plumage were watched at Knattabarrow Pool on 3 May 1942, although they could not be relocated on the 13 June. During fieldwork for the *Devon Tetrad Atlas* (Sitters 1988) two pairs, with trilling males, were found at Brockhill Mires, Dean Moor on 25 May 1981, and another registration was made for the area of Aune Head Mire. These records are very unusual, and no birds have been found there in recent years.

Breeding birds arrive in late April or early May, and leave the moor after breeding in July or early August. There have been two sightings of parties of birds that could have referred to the Dartmoor population moving off the moor, or passing migrants from elsewhere. Eleven flew northeast over Headland Warren on 12 August 1982, and seven flew southeast near Greencoombe, in the same general area, on 19 July 2003. Migrants, usually singles, have been seen in recent years at Cadover Bridge, Portworthy Dam, Fernworthy and Lee Moor, but they are by no means annual. Three at Knattabarrow Pool on 17 April 2000 were unusual. A most strange record, no doubt weather-related, concerned a flock of 50 feeding at Widecombe on 8 December 1987.

Dunlin

RUFF *Philomachus pugnax*
Rare passage migrant, and occasional winter visitor.
D'Urban and Mathew mention a male killed on Dartmoor in January 1867, and there was another record of one on the moor in October 1878 (Hurrell 1947). There is then a gap of over seventy years before the next birds were recorded on Yelverton Aerodrome on 26 February 1949, when three were present, with one remaining until the next day.

The locality with the most records is Meldon Reservoir. There, birds have been seen on autumn passage in four years. Two were recorded on 27 September 1972, an excellent count of six on 6 August 1975, and singles on 4 August 1976 and 20 August 1984. The only spring record has been two with Golden Plover at Willsworthy on 21 April 1992. A female was at Burrator on 25 August 1996, and the total for the moor is brought up to date with singles at Portworthy on 15 and 30 September 1997.

JACK SNIPE *Lymnocryptes minimus*
Very scarce winter visitor and passage migrant.
Due to their habit of sitting tight almost to the point of being trodden on, it is probable that many birds are missed and that they are rather more frequent on Dartmoor than records would suggest. As with Snipe and Woodcock, the true numbers in an area are more likely to be known by shooters than birdwatchers.

D'Urban and Mathew gave 10 September as the date when birds were expected back onto Dartmoor bogs in the autumn. They mention 13 shot at Raybarrow Mire one year on the 12 September. These dates are considerably earlier than birds have been seen on the moor in more recent autumns. Since 1930, the earliest date in the autumn has been 15 October. D'Urban and Mathew also stated that birds were known to linger longer on Dartmoor than elsewhere in south Devon, with some staying until March. Recent records have shown birds still being recorded up to the 19 April. Birds are largely absent from British moorland areas in winter (Lack 1986), so the small wintering population on Dartmoor is especially interesting.

By far the majority of records occur in the latter half of November and December. Birds are usually solitary, but can be found in small numbers in favoured habitat. The largest numbers have been on a private marshland site used for shooting near Okehampton. Eight were there on 20 November 2000, and six to 10 on 31 December 2001. Elsewhere at least five were flushed from a sewage field at Yelverton on 1 February 1947, and four were at Wigford Down on 6 December 2001.

In the past it has been suggested that, based on shooting bags, there could be one Jack Snipe for every wintering eight Common Snipe, but this estimate is now thought to be too high (Wernham et al 2002). The only comparison from a Dartmoor location was a post-Christmas shoot at Okehampton on 27 December 1975, that produced three Jacks in a bag of over 50 Snipe. If the ratio shown here of one Jack to about 17 Snipe is more correct, then there must be many Jack Snipe wintering on Dartmoor that are unaccounted for. Birds can turn up on almost any open ground with sufficient low cover, and are almost as likely to be flushed from a bracken-covered hillside as from a marshland area. One on burnt wet heath at Yarner Wood 31 January to 1 February 1994, was a first for the Reserve, and one particularly well-recorded bird was first flushed at Riddon Ridge on 30 January 1994, only to be flushed again from the same spot on 1 March and 13 April. Birds also favour the old mica dam tops of the clay pit areas around Cadover Bridge and Lee Moor, and are seen most years.

SNIPE *Gallinago gallinago*
Scarce migrant or resident breeder, winter visitor and passage migrant.
D'Urban and Mathew noted that Snipe bred in small numbers on Dartmoor. Birds have probably always been widespread breeders in suitable valley bogs, but in the early twentieth century there were few records to confirm this.

In 1933 they were noted as nesting in considerable numbers in Raybarrow Mire, a place where they are still found today. Four pairs bred on Sourton Down in 1934, with three pairs near Merripit Hill in 1939, two or three pairs at Cadover Bridge and two pairs at Postbridge the same year. There is then a lack of records in the *Devon Bird Reports*, suggesting a lack of observers rather than a lack of birds. As usual it is Peter Dare and Ivor Hamilton who come to the rescue with substantial records of birds breeding in the Postbridge area. Fifteen to 20 pairs bred in the area from 1956 to 1962, but they noted a considerable decrease after the bad winter of 1962/63 (Dare and Hamilton 1968). Nowadays eight to ten pairs are thought to nest at seven sites around Postbridge (Dare 1996).

The great problem in establishing numbers of breeding Snipe is their ability to remain quiet and out of sight during most of the daylight hours. The only way to make any estimate is to visit likely areas at dawn or dusk, when males will be in the air displaying and 'drumming', with perhaps others 'chipping' from cover. Surveys of the moor in the past have not allowed for this, and thus it is probable that totals obtained have been under-estimated. This has been fully appreciated by the organisers of the surveys, but to date no survey work aimed specifically at Snipe has been carried out. The 1979 RSPB/DNPA Survey recorded a total of 90 pairs. The RSPB Survey of 1992 recorded 10 territories in certain areas of the northern moor, where the same number had been found in 1979. However, a survey of the same area in 1986 had produced no birds. *The Dartmoor Moorland Breeding Bird Survey* of 2000 found no evidence that the population had changed since 1979. A figure of 100 to 200 pairs was estimated, which overlapped with the estimated figure of 110 to 120 in 1979. This figure for 1979 took into account the probable number present but not found, in addition to the 90 pairs confirmed. A caveat was given regarding the 2000 survey, that methods employed were not well suited to monitoring the species, for the reasons already given (Geary 2000).

Although no specific survey has been undertaken, counts by members of the Dartmoor Study Group on summer evenings have suggested that breeding numbers could exceed even the higher estimate of the 2000 Survey. Snipe are present in nearly every bog or mire of any size, and certain areas have a considerable number of territories. The Whiteworks/Fox Tor Mire area had about 20 birds present on 8 May 2000, with 10 to 15 'drumming' males. Langstone Mire, on Peter Tavy Great Common held a minimum of six birds on 27 April 2000, with several in May and June. At least 10 were 'chipping/drumming' at Halshanger Common Marsh in April and May 1999, with six at Blackslade Mire and four at Bagtor Down, both in May and June of the same year. Three were overhead 'drumming' and a further five 'chipping' in the bog during an evening visit to Raybarrow Pool on 22 June 1999. Despite certain local decreases, it would appear that the Snipe population of Dartmoor is very healthy and probably increasing. This is of great significance, as elsewhere in the Southwest, and Britain generally, a marked decrease has been noted in recent years.

After the breeding season birds probably stay on the moor, and numbers are then increased by birds from elsewhere in autumn and winter. If, however, severe weather hits and ground becomes frozen, the moorland bog sites are quickly abandoned for lower levels. Large winter congregations are not at all common, but certain sites will hold smaller numbers year after year. The old mica dam in the clay pit area of Cadover Bridge regularly has up to 30 most winters, especially if the temperature has been sufficiently low to harden the ground on surrounding moorland. Portworthy Dam also holds similar numbers in autumn and winter. The maximum count for Dartmoor, however, would appear to be the autumn record of about 100 at Postbridge on 6 October 1974.

GREAT SNIPE *G. media*
Very rare vagrant, with no records since the nineteenth century.
This species is included on the Dartmoor list purely on the basis of old records, accepted at a time when it was breeding over a larger area of Europe than at present, with very many more autumn records of vagrants occurring in this country. Since the nineteenth century there has been a marked contraction of their European breeding range, and a collapse in breeding numbers. Consequently Great Snipe has become a super rarity in Britain. With its rarity status has come more detailed analysis of its identification, and it is probable that certain old records would not be acceptable today.

However, the records given in D'Urban and Mathew have a ring of truth about them, and are as follows:

1846: One killed on Dartmoor in November.
1850: One on Shaugh Moor on 7 September.
1854: One at Ashburton on 7 October.
1868: One on Dartmoor on 29 September. This was the only bird that
      D'Urban and Mathew had encountered themselves on the moor.
      It was originally flushed a few days earlier on a dry hillside, and
      was found again on 29 in the same spot, when they succeeded
      in shooting it.

1876: One shot on Dartmoor on 23 August. This bird was claimed to
have been an adult, but D'Urban and Mathew viewed this with some
suspicion, as birds obtained in the autumn were invariably juveniles.
1886: One shot on Dartmoor in October.

WOODCOCK *Scolopax rusticola*
Winter visitor and probable passage migrant. Has bred in the past, but present breeding status unclear.

The breeding status of this woodland wader is difficult to assess. D'Urban and Mathew quote references to breeding in the nineteenth century, but nothing specific. The statement in the *Devon Bird Report* for 1964 that a nest with two eggs had been found at Meavy about 50 years previously, is the nearest that we come to a confirmed Dartmoor breeding record. There have been several pointers to breeding, though, that taken together indicate that birds may breed unobserved at times in Dartmoor woodland.

One, at Burrator on 3 May 1975, was at a time of year that suggested possible nesting, as was another at the same locality on 11 May 1996. Remains of prey at a Buzzard's nest at Moretonhampstead in 1966 included a young Woodcock. During the fieldwork for the *Devon Tetrad Atlas* (Sitters 1988), there were two registrations for possible breeding on the moor. In 1979 a bird was seen in May/June at Holming Beam Plantation, and on 12 August 1982 a freshly killed bird was found at Vitifer.

In autumn, migrants arrive from Scandinavia. Numbers depend on breeding success and weather conditions, with the main influx in late October. Birds can then be flushed in just about any woodland, although the conifer plantations record the majority. Because of their quiet, crepuscular habits Woodcock are frequently overlooked by birdwatchers. The numbers reported are only a fraction of those present. The only real indication of wintering numbers is through the records of organised shoots. Recent records have included 40 rises during a moor-edge shoot on southern Dartmoor on 17 January 2001, a total of 76 rises on two shoots in the Walkham Valley on 28 November and 1 December 2001, as well as 13 rises at Sourton Cross on 13 November 2002, and in excess of 30 on 9 December 2002 on a shoot between Princetown and Tavistock. Most of the wintering population departs during March, especially in the latter two weeks, although a few stay into April. Of course at this late date they could well be migrants passing through the region from further south. The only indication we have had regarding the origins of our wintering population comes from a bird ringed in woodland near Ashburton on 14 November 1987, and later shot at Hammarby, Orebro, Sweden on 3 July 1990.

Wintering birds usually spend the day in woodland cover and flight out to nearby fields or moor in the evening to feed, returning again by dawn.

Occasionally birds will be found on open moorland during the day, well away from any type of woodland. Records of this type have come from Ryders Hill on 24 November 1985, Okement Hill on 8 December 1998 and 16 January 2000, and Plym Head on 15 December 2001.

BLACK-TAILED GODWIT *Limosa limosa*
Very rare passage migrant.
Just one record of a bird that went off passage for a day at Lee Moor clay pits on 7 August 1995.

WHIMBREL *Numenius phaeopus*
Very scarce passage migrant.
Records of birds seen on or over Dartmoor are quite infrequent, but there are enough additional records of birds heard passing over at night to conclude that a small over-moor passage may be annual, especially in spring.

The earliest spring record is of a bird at Portworthy Dam on 9 April 2000, and the latest in autumn were birds passing over Okehampton at night on 29 August 1987. Most spring birds have occurred during the peak passage times of late April and mid-May. The autumn records have all been during August. There has been one mid-summer record of an individual flying over Headland Warren on 11 June 2001.

The only large flock seen on the moor was an unprecedented 50 resting on grass moor at Sourton Tors on 14 April 1947. They flew off to the north. The period of 5 to 17 May 1998 saw more than usual activity over the moor, with birds heard calling as they passed over on several nights, and two flocks of 10 being seen. The first was at Shorts Down on 5 May, where birds were feeding in short grass, and the second at Pew Tor where the flock flew over high going east, on the 14 May. Other numbers of note have been seven at Portworthy Dam on 19 April 1998, and eight there on 19 August 2000.

CURLEW *N. arquata*
Rare and declining migrant breeder, and occasional passage migrant.
Curlew traditionally arrive at breeding sites on the moor in early March. Once on territory, they are quite easily located by their frequent flights and evocative bubbling song. The actual nesting site can be much harder to find, as birds move around over a wide area. Early in the season pairs can be seen leaving the nesting area and flying to nearby sites to feed. Later in the season, off-duty birds can be watched flying off some way from the nest site. All this can add up to a rather confusing picture when pairs, sometimes in close proximity, have to be assessed as breeders or otherwise. After the breeding season birds quietly leave the moor in late June or July, and return to the estuaries.

Breeding Curlew have never been widespread over Dartmoor. They prefer marshes and wet rushy meadows, and have seldom occurred on high moorland. The central basin area between the northern and southern uplands, centred around Postbridge, has always been their stronghold, along with one or two similar areas mainly on the east of the moor. As the population has shown a recent marked decline, it would be perhaps helpful to look at these areas in more detail:

Postbridge area. The main area over the years, and because of survey work in the 1950s and 60s, the area of which we have the most detail. Records show that many pairs were in the area in the 1930s. There was then indications of decline in the 1940s, but the population was still a healthy20 to 25 pairs in the 1956/62 period. After the severe winter of 1962/63 breeding numbers decreased, and only returned to 50% of their former levels by 1968 (Dare and Hamilton 1968). The decline continued until by the early 1990s a maximum of only seven or eight pairs were thought to be in the area (Dare 1996). Even this may have been an over-estimate, as by the end of the decade only one or two pairs could be found. As of 2004 the picture is the same. The reasons for this decline are as yet not fully understood, but it has been noted that most of the pairs from the early 1990s onwards have nested in bog and marshland sites that are difficult to approach, and that the decline to some extent at least may have been related to public disturbance and increased stock grazing (Dare 1996).

Muddilake, Cherry Brook and West Dart areas: Lying to the south and south-west of the Postbridge area, this area was not included in the survey work of Dare and Hamilton. No early information is therefore available, but it has no doubt been a well-used and traditional area. Curlew were not specifically noted here until 1989, but four territories were found in 1992 and 1994. Since then numbers have fallen. The 1999 survey by the Dartmoor Study Group, and the DNPA 2000 survey could only find one pair.

Blackslade Mire, Rippon Tor, Bag Tor and Halshanger areas: Rather ironically, as it is surely one of the most visited, and thus disturbed, areas of the moor, this is now the main area for breeding Curlew on Dartmoor. There are few old records, but pairs have been noted since 1984, although as with other areas, they were probably breeding here at a much earlier date. Several marshes in the district have had breeding pairs, and it has not always been easy to pinpoint nest sites. However, at least two pairs were present in 1988 and 1989, with three to five pairs in 1993. During the survey work of 1999 and 2000, three pairs were thought to have been present each year, but further work in 2003 indicated that five or six pairs were then present, the only optimistic sign regarding this species at this time. Less encouraging was that during 2003 none of the pairs were thought to have bred successfully.

Riddon Ridge and Dury Farm areas: Birds have been seen here in summers since at least 1988. It would appear that this has been a regular site for a pair until 1999. A pair has not been located since.

Birch Tor, Warren House Inn, Boveycoombe Head, Hameldown areas: Displaying birds have been noted in this area in several years 1983 to 1995. It is probable that only one or possibly two pairs were ever present. A pair was seen on three occasions in June 2003 in the Challacombe Farm area.

In the years of the Devon Tetrad Atlas fieldwork from 1977 to 1985, there were several registrations, including a confirmed breeding, in the Mary Tavy/Black Down area. This area does not now appear to hold breeding birds, although birds were seen

Curlew

around Black Down in 1999. Other areas that have had occasional breeding pairs in the past have been Cadover Bridge, with birds noted in 1938 and 1996, Fernworthy with two pairs in 1950, Lydgate Buckfastleigh Moor, with a nesting pair in 1946, Sharp Tor, Lydford with a pair in 1993, and Scorhill Down, with a territory recorded in 1994. Rather more unusual, because of their higher altitude, were the pairs on the North Teign near Quintens Man, and Sittaford Tor (perhaps the same pair) in 1997, and Wildbanks Hill in 1995.

Professional surveys of the moor have only confirmed the reduction in numbers noted by casual observations. The RSPB/DNPA Survey of 1979 found a total of 23 pairs, in itself a greatly reduced number compared with the number that would have been recorded 20 years or so earlier. The DNPA Survey of 2000 found only four pairs.

Records outside the breeding season are exceptionally few. Occasionally birds moving south in the autumn are encountered, as the bird seen at Taw Marsh on 14 August 1982, and the three flying south at Sourton Cross on 8 October 2002. More interesting, however, is the record of a flock of 30 migrants resting on the moor at Butterdon Hill on 24 July 1938. The only winter records have been an individual at Fernworthy on 14 December 1968, four at Tor Royal on 27 November 1994, and one that flew over Horrabridge calling at 2200 11 February 2001.

REDSHANK *Tringa totanus*
Very scarce passage migrant.
Considering the large number of birds that favour the estuaries of Devon it is surprising how few have been seen on Dartmoor. There has only been one record in spring, and this involved a party of 14 that came off passage at Portworthy Dam on 17 April 1998. This is also the only flock seen in the Dartmoor area.

Of the remaining 20 records, have all but two have involved single birds. Most of the wanderers arrived in the last week of June or early July, suggesting that they were non-breeders or juveniles from British stock. Others in autumn were reported in August or early September and were probably migrants from further afield. Burrator, Fernworthy and Portworthy Dam have several records each, and Meldon and Kennick Reservoirs have each had one record.

GREENSHANK *T. nebularia*
Very scarce passage migrant, and very rare winter visitor.
The total number of Greenshanks recorded on Dartmoor up until the end of 2002 is 49. Spring passage has only accounted for four birds, and winter records are only for five. The rest are all autumn migrants, with a high proportion within a rather limited period.

There are two records in January, both from the china clay pit areas of Lee Moor and Portworthy Dam. A single was present on 29 January 1999, and two on 4 January 2002. One of the other winter records was also from Portworthy Dam, where a bird was seen on 16 November 1997. The remaining winter record was rather different as it involved a bird that dropped into a nature reserve pool at Throwleigh on 5 December 2001.

In spring an early migrant was at Cadover Bridge on 10 March 1997, and singles were at Portworthy Dam on 8 May 1997, and 4 May 2000. An additional record was of a bird over-flying the moor at Soussons, going north on 11 May 1968.

The earliest on autumn passage was at Fernworthy on 17 July 1995. The majority of autumn migrants passed through, usually singly, during August, with about 55% in the last week of the month. Mainly reservoir sites are visited with both Burrator and Fernworthy having many records each. Other reservoirs and clay pit areas have also had visits. After the end of August migrants are few, with only twelve occurring in September. The latest autumn record was of a bird flying south over Princetown on 29 September 1965. The maximum number together on autumn passage has been four, seen at Burrator on 27 August 1955.

GREEN SANDPIPER *T. ochropus*
Scarce passage migrant and very rare winter visitor.
The occurrence of this species is very similar to Greenshank, except that there have been rather more records, and the peak of the autumn passage is a little earlier. There are on record a total of about 115 birds. A maximum of eight have been in spring, two in winter, and the remainder during the rather protracted autumn passage from

June to October.

The two winter records, both at the end of December, involved a bird feeding in a sewage field at Yelverton on 30 December 1945, and one on the West Dart on 26 December 1963. The Yelverton bird wintered and was seen again in January and February 1946. Spring migrants were three on the River Meavy at Clearbrook on 1 March 1947, up to two at Sourton Marsh 28 March 1951, one at Shipley Bridge 11 April 1946, and two again at Sourton Marsh 22 April 1951.

The earliest bird in autumn was one at Portworthy Dam on 12 June 2000. There are a few more records for June and early July, then the beginnings of a build-up in late July. August is the main passage month with 75 of the total records, and 65 (87%) of these occurring between 6 and 24 August. Only 12 birds have been recorded in September, and the only October record was of one or two at Sourton Marsh on 8 October 1951.

Most records are of ones and twos, but at times larger groups are seen. Four were together at Burrator on 18 August 1977, and at Meldon Reservoir from 6 to 15 August 1983. Five were at Meldon 14 August 1977, six at Drakelands, Crownhill Down on 31 July 1999, and a record eight at Meldon again on 8 August 1977. As would be expected most of the passage birds have been at reservoir sites. But this is a species that will stop off at any patch of water no matter how small. Farm and garden ponds are quite acceptable for a quick feed until disturbed, and even swimming pools have attracted birds. Moorland streams are also enough to entice birds down, as with one below Warren House Inn on 20 April 1965. Two interesting records from the high moor involved one at Hangingstone Hill on 4 August 1947, and one west of Taw Head on 19 June 2000.

WOOD SANDPIPER *T. glareola*
Rare passage migrant.

There are only nine records of this sandpiper, which is always scarce in the west of England. Two of the birds have been in spring, and the remainder in autumn between 23 June and 20 August. Records have been more frequent in recent years.

1971: Meldon Reservoir, 26 April, one.
1994: Lee Moor Clay Pits, 11 August, one.
1995: Smallhanger, Crownhill Down, 10 August, one.
1999: A good year nationally for autumn migrants, that was reflected in
    Devon and on Dartmoor.
    Burrator, 25 July, one.
    Portworthy Dam, 1 August, one.
    Venford Reservoir, 14 August, one.
2000: Fernworthy, 20 August, one.
2001: Smallhanger, Crownhill Down, 23 June, one.
2002: Conies Down, 18 May, one flying around over the top of the
    Down, calling frequently.

COMMON SANDPIPER *Actitis hypoleucos*
Scarce passage migrant and very rare migrant breeder. Very rare winter visitor.

Writing at the end of the nineteenth century, D'Urban and Mathew knew the Common Sandpiper to breed on the Okement, Avon and other Dartmoor streams, and they frequently found nests. They no doubt continued to be regular through the early years of the twentieth century, although records are largely lacking. The early *Devon Bird Reports* for 1929 and 1933 mention that pairs nested frequently in favourite localities. By the early 1950s, things were begining to change, and what had been an accepted and widespread breeding bird on Dartmoor streams began to get very much rarer. This decline is best looked at by locations:

River Plym, Cadover Bridge, and latterly Lee Moor Clay Pits, Portworthy Dam and Crownhill Down. There were records for this area from at least 1933, and it would look almost certain that nesting occurred well before that but got little attention. By the early 1950s the area was apparently deserted (Moore 1969). Displaying birds were seen again in May 1961, but this was the last record until a breeding pair was located near Cadover Bridge in 1977. They bred successfully. It had always been this stretch of the Plym that had held pairs, although a pair was found near the source in 1937. After 1977 birds were noted most years, although proof of breeding was hard to confirm. A pair certainly bred successfully in 1983 and 1989. In 1981 birds were watched in the summer around the mica dam of a nearby clay pit, and in the next few years birds increasingly took to this artificial habitat. They were still being seen on the river near Cadover in the late 1980s, but these could have been individuals from clay pit sites calling into feed. The maximum number breeding on china clay pit sites is difficult to establish, but by 1992 pits at Crownhill Down and Portworthy Dam had also had probable breeding pairs. Breeding was confirmed at Lee Moor in 1994, and in 1995 the combined pit areas probably had three pairs. Since then up to three pairs have been present most years, although proof of successful breeding has usually been lacking. It is probable that human disturbance played a major part in the bird's desertion of traditional river habitat. The only breeding on Bodmin Moor in recent years occurred in similar habitat on the edge of a china clay pit area.

Postbridge area: A pair was noted on the East Dart at Bellever Bridge on 22 April 1948. A pair nested successfully in the area the next year. Until about 1954 there were up to five pairs along the East Dart between Bellever Bridge and Postbridge, but gradually the numbers decreased until by 1959 only one successful pair was present. A pair was seen in 1960 and 1961 but there was no proof of breeding, and no birds have been seen since. The extinction of breeding birds in this area, as on the Plym, has been put down to increased human disturbance from visitors (Dare and Hamilton 1968).

Other sites used for breeding were occupied more sporadically, sometimes for a year or two only. A pair bred near Avon Dam in 1959, and again in 1962. On the River Swincombe a pair, possibly nesting, was found in June 1962 (P.J. Dare *pers.*

comm. 2004*).* At Meldon Pond a pair bred in 1968/69 (D. Bubear *pers.comm.* 2004*),* and at a northern moor location, probably Meldon Reservoir, a pair bred successfully in 1979, and may have nested in 1989. A pair nested on the West Dart at Sherberton in 1939, and a pair stayed the summer at Fernworthy Reservoir in 1948, but no nest or young were seen. Two pairs were seen on the Upper Teign/ Wallabrook on 4 July 1940, but although breeding was suspected it was not proved, and there were no other records for the area.

On spring passage the earliest record is of a bird at Cadover Bridge on 10 March 2000. This is so far in advance of the normal passage time that it may well have been a bird that wintered close by. The main passage begins in mid-April, with a peak during the last ten days of the month. Relatively few are seen in early May, and by June the birds seen are quite possibly non-breeders returning south. Autumn passage is protracted lasting from early July until the third week of September, with 44% of the autumn passage occurring in August. There has been only one October record, and a single record for November. This was a bird at Burrator on 13 November 1983, and could well have been a wintering bird stopping off on the way to some local estuary. There have been three winter records. A bird was at Portworthy Dam in January of 1994, and another was there in January and February of 1995. The third record was a bird at Burrator on 8 December 2003.

Numbers at sites in the spring tend to be low, so 13 at Burrator in April 1994 and 1995 was exceptional. The highest autumn counts have been ten at Fernworthy on 23 August 1996, and again on 18 July 1997, and 10 at Trenchford Reservoir on 5 July 1993.

TURNSTONE *Arenaria interpres*
Very rare passage migrant, and winter visitor.
On 28 November 1953 two observers must have been very surprised to find a Turnstone resting on open moor between Lynch Tor and White Tor. Perhaps it had been forced down by adverse weather, as they were able to approach it within five yards, suggesting that it was tired. The only other records have been of two at Portworthy Dam on 12 February 1995, and a single there on 7 October 2001.

GREY PHALAROPE *Phalaropus fulicarius*
Very rare autumn passage migrant.
The few Dartmoor records have all been associated with severe autumnal gales, and the subsequent displacement of birds inland.

A bird was found at Wrangaton in 1943, and the remains of another was picked up close by on the moor in mid-September 1950. One found swimming on a nature reserve pool at Throwleigh at dusk on 1 November 1994 was probably this species, although because of the poor light specific identity could not be claimed. The only other record was of a bird at Portworthy Dam on 7 October 2001.

ARCTIC SKUA *Stercorarius parasiticus*
Very rare passage migrant or accidental visitor.

Birds have occurred on four occasions. Most of these are likely to have been associated with severe gales blowing them inland whilst on migration.

D'Urban and Mathew were shown an adult that had been shot at Christow on 10 October 1873. On the 16 September 1959 a dead immature was picked up at Scorhill near Gidleigh. There was then a period of over 20 before the next record, which was also the only spring record. This record concerned a light phase bird that flew southwest over Brent Moor on 23 May 1981. The records were completed with a dark phase bird that flew west over Okehampton on 28 September 1982.

LONG-TAILED SKUA *S. longicaudus*
Very rare passage migrant or accidental visitor.

On 11 June 1961, a day of good visibility but cloudy with a chilly northwest wind, Peter Dare was watching Fernworthy Plantation from a vantage point between White Ridge and Grey Wethers, looking for the Montagu's Harriers that were present in the area. A long-winged, raptor-like bird was spotted approaching from the southeast over the skyline of Assycombe Hill. It flew past at quite close range, steadily beating into the wind towards Teignhead Farm at about tree top height. He was startled by its shape, and realised it was an adult Long Tailed Skua in full breeding plumage. With typical small structure, long, slender angled wings, and extraordinary foot-long central tail streamers undulating in the air-flow, it was unmistakable (P.J. Dare *pers. comm.* 2004) Surely one of the most awe inspiring of all Dartmoor records.

Overland migration by this species has been well documented (Cramp *et al* 1983), and no doubt this adult belonged to one of the far northern breeding populations that arrive at their nesting sites in late May and June. This goes some way to explain a high Arctic breeder over Dartmoor in summer, although it must have been a tremendous surprise to the lucky observer.

SKUA sp. *Stercorarius sp.*

A small, adult pale-phase skua was watched feeding on insects taken from the ground at Piles Hill on 22 and 23 September 1990. The observers considered it an Arctic Skua, because of its lack of tail extension, but from its behaviour of feeding on invertebrates the Records Committee of the *Devon Bird Report* thought it more likely to have been a Long-tailed Skua. They also pointed out that adult Long-tailed Skuas usually lack the central tail extension in September.

MEDITERRANEAN GULL *Larus melanocephalus*
Very rare passage migrant and winter visitor.

There are only four records for this gull. It has increased greatly on the coast in recent years, but is still a rarity inland.

A single was seen at Portworthy Dam in August 1993, and a summer-plumaged adult was there on 11 July 1998. Both of these records could have been of birds relocating from breeding sites in southern or eastern Britain, or from the continent. The remaining records are in winter and could relate to wanderers, or birds forced inland by weather conditions. A bird in second winter plumage was on Roborough Down on 3 January 2000, and an adult in summer plumage was at Burrator on 11 February 2001.

LAUGHING GULL *L. atricilla*
Very rare vagrant.
Between the 14 and 20 September 1996 an adult in non breeding plumage visited Portworthy Dam, with other *Larus* species, to roost. It spent the day around the Plym Estuary. This is Dartmoor's only record of this North American species.

BLACK-HEADED GULL *L. ridibundus*
Passage migrant and winter visitor.
Black-headed Gulls do not breed on Dartmoor. A few non-breeding birds are seen from April to June, but it is July before the birds are seen regularly during the post breeding southwesterly passage. In the early days of the Devon Bird Watching and Preservation Society, efforts were made to monitor this passage, but after the 1930s little comment was made in the annual reports, and it must be assumed that as the species was getting more regular at inland sites, interest in passage birds inland had tended to wain.

Around Dartmoor small but increasing numbers are noted through August and September, but numbers increase greatly in October and especially November. November is also the month of highest counts on some of Devon's estuaries. December and January counts on the moor are normally less than half of the November totals, and by February there is a marked fall in numbers again, as birds begin to drift back eastwards. By March birds are very scarce around the moor.

The main locality for Black-headed Gulls at any time of year is the moor edge, china clay area of Portworthy Dam. The highest counts there have been 600 in October 1994, and 350 in November 1996. Elsewhere numbers are scattered and usually small. Higher numbers can be linked with roosting migrants, or birds moving in response to adverse weather. The count of 570 at Trenchford Reservoir on 5 August 2001 was an example of the former, and 110 at Burrator on 5 February 1997, an example of the latter. A flock of 156 put up by Buzzards on Riddon Ridge on 8 January 1983, could well have been weather-related.

There have been two records of ringed birds that throw some light on the origins of our wintering population. An immature seen at Tavistock (possibly just outside the Dartmoor area) in 1965 had been ringed at Jokijarvi, Finland. It returned as an adult in both winter periods of 1966. A bird seen on Roborough Down on 4 January 1972 had been ringed on 12 June 1969 near Hilversum, Holland. It returned to the same spot again in the winters of 1973 and 1974.

COMMON GULL *L. canus*
Winter visitor and passage migrant.

The first autumn migrants to reach our area are seen occasionally in late July. They are probably birds from the northern British breeding colonies. The earliest was one at Lee Moor on 17 July 1986. Numbers remain low throughout the autumn, and it is only in November that an increase takes place. The increase builds during December, and the peak of wintering birds occurs during January. This is probably due to continental birds that have arrived earlier on the East Coast to winter, moving gradually westward as the season progresses (Wernham *et al* 2002). Numbers decrease rapidly in February, as birds move on westward, or perhaps move back eastward prior to spring migration. March records are very few, with 74 at Portworthy Dam in mid-March 1998 being the only record of note, and the only indication of any real spring migration over the moor.

As with Black-headed Gull, and in fact most gull species, Portworthy Dam is the only regular area. Indeed, after the Exe, it is probably the most important site in Devon for Common Gull. Maximum counts here have been 357 in November 1999, 257 in December 1999, 229 on 16 January 2000, and 330 in January 2001. Elsewhere records show little pattern, and could well be weather-related, although the highest numbers tend to be in the period of January movement. Burrator, not a site normally known for gulls, has had two big flocks. The first was over 300, all but one adults, on 2 January 1972, and the second was 325 on 19 January 1981. Other winter records of note have been 100 at Burrator on 7 December 1975, and the same number there on 6 December 1978, a maximum of 74 at Fernworthy Reservoir on 13 December 1987, 106 at Okehampton during February 1989, and 100 at Chagford Common on 9 January 1994.

LESSER BLACK-BACKED GULL *L. fuscus*
Passage migrant and winter visitor.

Fifty years ago the British population of Lesser Black-backed Gulls was almost totally migratory. Birds left in the autumn, to winter on the coasts of France and Spain, and returned to breed in the early spring. Since then a change has taken place in their habits and migration pattern. They have now taken advantage of inland feeding and roosting sites in Britain, and autumn migration brings many hundreds to sites in the Southwest, where once they would have been almost unknown. The timing of return migration has changed too. Birds do not stay away from Britain for so long, many returning during the winter, and indeed some probably not leaving Britain at all, and spending the winter in the South and Southwest (Wernham *et al* 2002). There are still many questions to be answered regarding the changed status of this species, but below is a brief background to the somewhat confusing picture emerging from recent Dartmoor records.

A very high proportion of Dartmoor numbers, about 80-90%, are recorded at Portworthy Dam. It is a site of national importance for the species, and is second only in importance in Devon to Roadford Reservoir. Early migrants start appearing here in June. Numbers are then usually quite low, but occasionally substantial numbers are present, as in 1995 and 1998 when the monthly maxima were 300 and 500 respectively. Large numbers at this time of year are no doubt non-breeders, wandering southwards ahead of the main autumn passage. An increase is noted in July, but it is in August and September that the passage peaks. Monthly maxima then quite regularly reach four figures, with the highest counts being 1500 in August 1995 and August 1996, and 2250 in September 1996. It is interesting to note that a particularly poor autumn for Portworthy Dam in 1998 coincided with the highest ever counts at Roadford Reservoir. October figures are substantially lower than the previous two months, but an increase is sometimes noted in November. An impressive 2000 were present in November 2001. Whether these represented late autumn migrants, wintering birds, or birds returning early from the continent is at present unknown. December and January records usually give quite low numbers, although 300 were counted in December 1998. The early months of the protracted spring migration bring very few birds to Portworthy. February brings only single figure counts, with negligible totals in March and April. May sometimes has a small upsurge in numbers, perhaps due to late migrants of the southern Scandinavian race *L.f. intermedius* being present, although none appear to have been specifically identified.

Elsewhere on the moor, with very much smaller numbers, the pattern is similar but a little more complex. There have been a few records for June, but rather more for July, including some interesting records of flocks feeding in freshly cut silage fields on the moor edge. These have included birds noted at Welltown, Walkhampton in 1997 and 1998, about 300 immatures feeding in fields at Nodden Gate during the morning of 6 July 2000, and 79 at Higher Godsworthy on 25 July 2001. The Nodden Gate record, and perhaps the others refer to non-breeders. Passage gets underway in earnest in August. Flocks are still to be found feeding in suitable fields, as with the 76 at Yellowmead, Sheepstor on 8 August 2002, but there is an increasing number of records of flocks over-flying the moor on passage. Several hundreds were noted going west at Prewley from 18 August 1970. Nearby Okehampton has seen passage in several years since 1985, with flocks moving between south and northwest. On 3 September 1985, 89 flew south in five minutes, just part of a heavy evening passage. Numbers fall sharply in October, although by the end of the month roosting flocks are beginning to build up at Fernworthy Reservoir. Records of a roost here go back to the late 1980s, but it was the late 1990s before slightly higher counts were being made. On 28 October 1997 37 were present, with 42 on 25 October 1998. On 28 October 2000 a count of 98 was made. All these pale into insignificance compared to the 500 or more roosting on 25 November 2002. Very few birds are around Dartmoor in December and January, but by mid-February the start of the spring migration has begun. Birds move on a broad front over the moor, and can be seen, singly or in small groups, almost anywhere until well into May.

Individuals of the dark-mantled southern Scandinavian race *L.f. intermedius* have been recorded on several occasions.One was at Fernworthy on 16 December 1974, two were at Okehampton on 10 January 1989, with four there on 14 November 1990. Single spring migrants have been noted at Langstone Moor on 18 May 1991, and at Venford Reservoir on the same date in 1993.

YELLOW-LEGGED GULL *L. michahellis*
Rare passage migrant.

This species, or perhaps more correctly sub-species, has only been seen at Portworthy Dam, but it is quite possible that individuals pass through elsewhere in the late summer and are not detected.

At Portworthy Dam, the first birds were identified in the gull roost in July 1994. After four adults on the 15 July, 20 birds of various ages were counted on 20 July, and 21 on the 22 August. Four adults were present on the 25 September, and a single adult on the 24 October. These numbers immediately placed Portworthy as the major site in Devon for birds, together with the Plym Estuary. There was thought to be an interchange of birds between these two sites.

Since then birds have been seen every year, up until at least 2000. Most have occurred in July and August. Numbers have never been so great as in the first year, with the maximum being 16 birds seen on 23 July 1998. There appears to have been a decline in numbers recently, with only singles being seen on three dates in 2000, and none in 2001.

HERRING GULL *L. argentatus*
Passage migrant and winter visitor, although status less than certain due to the large numbers on the coast that can be forced inland at times of severe weather.

With high numbers breeding around the coast, and some inland, it will always be difficult to be sure of the Herring Gull's true status on Dartmoor. What we may take to be passage may in truth only be feeding flights of coastal residents or birds forced inland by the weather. However, the records from Portworthy Dam, the only major site in the area, do show a pattern that bears similarities with the migrant Lesser Black-backed Gulls.

Numbers at Portworthy can increase sharply in June, perhaps swelled by non-breeders or locally-bred young. Peak counts occur here in August and September. Whether these are migrants or birds dispersing from coastal colonies, is unclear. October's maxima are normally much lower, and there appears to be no increase in November, unlike the rise some years in the number of Lesser Black-backed Gulls. Numbers are fairly low from December to February, and there is no sign of any spring movements in March or April, although numbers are a little higher in May. The highest counts at Portworthy Dam have been 1200 in June 1995 and 2000 in September 2001.

Birds can move over the moor at almost any time of year. Their flight paths and directions can appear rather aimless, and are thus difficult to assess. The majority could well be parties or families on feeding flights. Small flocks seen in moor-edge fields quite regularly, would suggest that this is the case. River valleys may well act as flyways for these flocks. Certain favoured areas, like the Prison Fields at Princetown, attract parties of varying sizes throughout the year. As with the Lesser Black-backed Gull, newly cut silage fields prove very attractive to birds in June and July. At times in June large flocks, made up almost entirely of immature birds, are encountered. In 1996 about 140 were noted in a cut grass field at Owley on the 30 June. About 400 immatures were present at Lee Moor on 3 June 2000, and a large influx of about 310 was seen at Fernworthy on 17 June 2002. Birds were feeding in nearby fields and on open moorland, as well as being present on the reservoir. They were all gone the next day. A similar, but much larger influx of about 750 was in the Fernworthy area on 7 June 2004. This influx was of special interest because, contrary to the usual pattern, it was made up of birds of all ages. It must be assumed that the proportion of adults present were non breeders, as it is known that up to 40% of birds with breeding experience do not breed in any one season (Wernham *et al* 2002).

The only reservoir that has acted as a roost site with any regularity over the years, is Fernworthy. Five hundred were estimated roosting on 14 December 1968, and 200 on 7 December 1975. In recent years numbers have been much lower, and roosting sporadic. The maximum was 84 on 11 November 1988. Away from Fernworthy the only record of note was of a very impressive 1510 that roosted at Trenchford Reservoir on 5 August 2001.

GLAUCOUS GULL *L. hyperboreus*
Very rare passage migrant.
A first winter bird was at Portworthy Dam on 23 April 1998, and another immature was there on 21 April 2000.

The first bird was the individual that had been present on the nearby Plym Estuary all winter and had joined the gull roost at Portworthy. The April 2000 record does not coincide with any bird on the Plym, and would have been a migrant from elsewhere moving north.

GREAT BLACK-BACKED GULL *L. marinus*
Scarce visitor from the coast and possible passage migrant. Has bred.
As its Latin name suggests, the Great Black-backed Gull is very largely a bird of the coast at all times of the year. It is also a species prone to get storm-blown in the mid-winter gales, and this time of year accounts for over half of Dartmoor records.

There have been 24 moorland records involving about 99 birds. As mentioned, the winter months of December to February account for 13 of these records involving 65 birds. Most records are of single birds, often seen at times of high winds. Occasionally a group will be seen, as with the seven at Fernworthy on 13 December 1989, and once a movement was noted with 44 passing south over Okehampton on 14 December

1986. Whatever the time of year, when noted in flight the direction has always been between west and south.

In spring, a rather puzzling movement of 23 going southwest in an hour was noted over Okehampton on 23 April 1988. Nine records involving 12 birds exist for the May to July period. Given the species known taste for carrion, it is possible that these records represent feeding forays by local coastal breeders. Instances of birds feeding on carrion have been recorded on several occasions. On 7 July 1946 at Ugborough Beacon, one on a sheep carcass made six Ravens keep their distance, and at the same locality on 17 January 1949, three birds with 13 Ravens cleared the carcass of a sheep in two days. Three were seen feeding on a carcass at Teignhead Farm on 8 July 1948. At times non-breeders will stay around an area for lengthly periods in the breeding season, as a bird did at Lee Moor from 1 April to 30 June 2002. A pair stayed for several months at Burrator Reservoir in the summer of 2003, a time in which rather more than usual were being seen around the moor.

In 1998, a totally unexpected and unprecedented event occurred when a pair bred successfully on a flat roof at Buckfast Abbey. This was the first inland breeding in Devon. The birds bred again in 1999, raising two young, and again in 2000 raising one young.

KITTIWAKE *Rissa tridactyla*
Accidental visitor.
There have been five records of this maritime species on Dartmoor. One has been in winter, and the other four have been in spring. All but one have been directly associated with violent gales.

In 1950 an adult was watched struggling west into the moor at Wrangaton on 10 April, amidst rain and strong winds. At Shipley Bridge on 7 April 1954, the remains of one was picked up, no doubt the victim of similar bad weather. The only record that may not have been attributable directly to the weather was a first summer bird moving north near Princetown on 9 May 1992. Another first summer individual was seen struggling against the weather at Dean Prior on 1 April 1994, and a freshly dead adult was found at Burrator on 2 January 1998, after severe southwesterly gales the previous day.

COMMON TERN *Sterna hirundo*
Very rare passage migrant.
The one spring record is of four birds at Two Bridges on 29 April 1995. In autumn two flew south at Wrangaton in the evening of 25 September 1954. Another record showing how weather can affect seabird migration occurred on 24 August 1999. It was a day of force 5 southeasterly winds, mist and drizzle. Observers all around the Devon coasts were noting a large passage of Common Terns in what turned out to be the biggest movement for many years. Some obviously were disorientated by the conditions and poor visibility, as 18 turned up at Portworthy Dam, raising the question of how many more crossed the moor undetected.

ARCTIC TERN *S. paradisaea*
Very rare passage migrant.
The few confirmed records have all come from the china clay district of Lee Moor or Portworthy Dam. A juvenile was seen on 18 September 1993, and another bird stayed for a few days from 30 August to 4 September 1995. A bird was seen on 1 September 1997, and what was either this species or Common Tern was seen briefly on 7 August 2001.

LITTLE TERN *S. albifrons*
Very rare passage migrant.
In recent years coastal records of Little Terns on passage have declined greatly. Inland records have always been extremely rare, and so a spring migrant at Portworthy Dam on 31 May 1998 was exceptional.

BLACK TERN *Chlidonias niger*
Rare passage migrant.
There have been nine records of this marsh tern occurring on Dartmoor, involving some 20 birds. This species is seen inland regularly in small numbers on migration, so the Dartmoor total appears quite low.

There has only been one spring record of two birds at Cadover Bridge on 13 April 1994. The first of the autumn records was a juvenile at Lee Moor on 10 August 1995. September has most of the autumn migrants, with two at Burrator 16 September 1970, one at Portworthy Dam on 21 September 1997, and one at Burrator on 22 August 1980. The only party to be seen over the moor occurred on 22 September 1957, when nine birds flew upstream the full length of Fernworthy Reservoir, then flew on out of sight westwards. An influx was noted at other sites in Devon around this time.

There are three October sightings. A single was at Cox Tor on 1 October 1983, two were at Portworthy Dam on 1 October 1995, and the remains of a bird was picked up in a field at South Brent on the very late date of 27 October 1952. The Cox Tor record was especially interesting as the bird was watched in typical feeding flight, hawking craneflies over the flooded car park on the B3357.

LITTLE AUK *Alle alle*
Accidental visitor.
Although none actually on the moor, there have been three records of this storm-blown waif on the borders. At a time when northwesterly gales stranded many in Devon, one was picked up on the road at Buckfast on 15 February 1950. Another was found at Ashburton on 1 December 1954, and a third was found in a lane near North Bovey on 24 November 1991. The three dates are typical of the period when Little Auks get "wrecked" on our shores, and also typical, unfortunately, was the fact that all three died shortly after being found.

STOCK DOVE *Columba oenas*
Resident breeder, winter visitor and passage migrant.

Writing in the late nineteenth century D'Urban and Mathew found Stock Doves to be spreading in Devon, but still not widespread. Wintering flocks were known, especially in the east of the county.

There were no early references to Dartmoor, and it was only in the 1950s that any assessment of breeding numbers was made. In the Postbridge area 15 to 20 pairs bred in the period 1956 to 1967 (Dare and Hamilton 1968). They nested mainly in the old mine gullies at Vitifer, although one or two pairs nested in ruined buildings at Powder Mills, and there were one or two tree nesters. A reappraisal of the situation in the early 1990s found 10 to 15 pairs, which were entirely arboreal, including some in Soussons spruce plantation (Dare 1996). Since then one or two pairs have been found back in Vitifer gullies.

The coverage of a 76.7 hectare Common Bird Census plot at Harford has shown territories since its beginnings in 1988 to at least 2000. Five pairs/territories were found in 1988, and the total has been between four and six for most years since. The poorest year, with only three pairs, was 1993, and the best years were 1997, with seven pairs, and 2000, with nine. Elsewhere, although no constant effort sites have been involved, the picture emerges of a species that probably inhabits most suitable areas but at low densities, so their presence is not always apparent. At Yarner Wood a pair nested in a box erected for Tawny Owls from 1980 to 1989, but not apparently since. It is this species's willingness to accept man-made structures for nesting that make it particularly interesting. Ruined buildings and old barns are used regularly, and in recent years this has been noted in the West Dart valley at Moorlands Farm, Prince Hall and nearby at Tor Royal. Barn sites were also being used at Bellever in 2000 and 2001, and at Middle Stoke, Holne in 2002. The bulk of the population still no doubt nests in traditional tree cavities, and are spread around most moor-edge and wooded valley sites. Most tend to be single pair sites, easily overlooked unless birds are calling, but three pairs were found at Two Bridges in 1990, and two at Sticklepath in 1996. In addition to the Soussons area mentioned above, birds have shown a tendency to take up territories in other conifer plantations at Burrator, Fernworthy, and Bellever.

In winter, the resident population is swelled by an influx of birds from the east. As with Wood Pigeon, this influx usually begins in November, and if flocks find sufficient food, numbers will build to a peak in January, with most birds departing again by March. Large gatherings have included 136 in the West Webburn valley on 4 March 1957, 200 or more at Hennock during January 1983, and 50 at Bridford on 13 December 2000, which increased to 250 by 6 January 2001. Occasionally visible migration can be seen, as with the 170 that moved southeast over Okehampton on 6 November 1989.

WOODPIGEON *C. palumbus*
Resident breeder, passage migrant and winter visitor.

This has always been a common species in Devon, and indications are that it has always been regular around the wooded edges of Dartmoor, although absent from the largely treeless interior. The planting of conifer plantations in the early twentieth century allowed birds to move into new areas. Breeding numbers in plantations have prospered over the years.

In the plantations of the Postbridge area they were noted as a common resident in the 1956 to 1967 period (Dare and Hamilton 1968). Breeding numbers were apparently unaffected by the severe 1962/63 winter. Doubtless the population left the moor and found sustenance at lower levels. The breeding population of the area at this time was estimated to be 75-150 pairs. Elsewhere estimates of breeding numbers have been rather limited. In the Harford farmland Common Bird Census plot of 76.7 hectares estimates between 1987 and 2001 have averaged 20 pairs, with a wide variation between the high of 31 in 1989, and the low of only 14 in 1993. Six territories were located in a tetrad on the West Dart, north of Hexworthy, on 11 May 2002, and 12 territories were found in a tetrad at Bonehill, northeast of Widecombe, the next day. Survey work on Throwleigh Common found 10 pairs/territories in a one kilometer square on 14 May 2000, with seven relocated on 10 June.

The breeding season for Woodpigeon is very protracted, and this has been reflected in certain Dartmoor records. In 2001 the first song and display was noted at Huccaby on 6 January. A late nest with two newly fledged young at Shaugh Prior was recorded on 13 October 1986, and two recently fledged young were seen at Brisworthy on 14 October 2002.

In late spring and early summer, birds can be seen on some days on flight lines taking them towards the moor. Often river valleys are followed. These movements are thought to be purely for feeding. In June and July birds fly onto the open moor for whortleberries, and parties have been noted at Yes Tor and Arms Tor in recent times.

In late autumn there is an annual passage of birds south or west across the moor. This is part of a movement that is seen over much of southern England mainly on the coast. Ringing returns have shown that British Woodpigeons are mainly sedentary, so it would seem logical that these autumn flocks are of continental origin. However, there is little evidence of continental birds being found in this country. In fact, up until 2002 only eight continental ringed birds had been found in Britain (Wernham *et al* 2002). It would seem probable that the autumn flocks arrive on the east coast from Scandinavia and move rapidly south and west through Britain, *en route* to France. Their movement is so fast, and in such large numbers that the limited ringing data available is of little use in establishing their origin (Wernham *et al* 2002). Flocks are seen every year over Dartmoor from mid-October. The earliest could be local bird movements, but by early November the movement is more pronounced, and no doubt migrants are involved. This passage has been noted since at least 1938, and can be witnessed almost anywhere on the moor, although the highest numbers have been seen passing around the moor-edge, or west along river valley flyways. The highest totals have been approximately 700 moving south at Wrangaton before breakfast on 31 October 1945, 530 moving south-west over Vitifer on 2 November 1975, 570 flying west around the southern edge of the moor at Western Beacon on 4 November 2002, and an amazing 4000 flying southeast over Okehampton, between 7.30 am and 8.30 am on 14 November 1989.

As well as overhead passage, feeding flocks are noted throughout the autumn and winter. Some of the feeding flocks in autumn may consist of migrants off passage. During the winter, roosts build up in the conifer plantations. At Soussons on 12 January 1983, 450 were recorded roosting, and rather smaller numbers have been seen at Burrator, Fernworthy and Clearbrook in recent years. Birds numbering about 300 flew over Huccaby to roost on 16 December 2000, and 360 were watched entering a roost at nearby Brimpts Plantation on 29 January 2001, with 350 the next night.

COLLARED DOVE *Streptopelia decaocto*
Resident breeder.

Collared Doves first colonised Britain from the continent in 1955, and the first birds in Devon were noted in Plymouth city centre in 1960 (Moore 1969).

On Dartmoor, two visited a poultry yard in Postbridge in mid-May 1966. They were still in the area in June, but did not breed. The colonisation of Dartmoor has been slow and gradual over the years since then. In the remaining years of the 1960s, Okehampton was the only town to record breeding birds. Successful breeding first occurred there in 1968, and the population was noted to be increasing by the early 1970s. However, in 1994 and again in 2001, only three pairs were recorded, suggesting that in the years since the 1970s they had not fared too well.

The 1970s produced the first birds at Ivybridge in 1971, a rather high total of six pairs in Bovey Tracey in 1972, and birds at Haytor, Ilsington, and Prewley in 1973. The first birds were seen at Chagford in 1975, and breeding was proven at Cornwood in 1976, although the pair did not stay to breed in 1977. Dousland was first occupied in 1979, with good numbers about in the area ever since. Up to 17 birds were noted here in 1982. Interestingly, Postbridge which hosted the first arrivals in 1966, did not have a confirmed breeding until the 1980s (Dare 1996).

Grenofen had birds throughout the year in 1982, with small numbers until at least 1994. The first were seen at Buckfastleigh in the summer of 1983, and 10 were in a garden at Crapstone on 5 October 1986. The 1990s brought birds to many of the villages and outlying areas away from towns, and into several areas quite far into the moor. Throwleigh had birds for the first time in 1990, and a single was seen at Burrator in 1991. Welltown, Walkhampton had records in 1994, as did Lynch Common near Meavy. A pair bred at Sticklepath in 1995, and pairs have bred there ever since with three in 1999. Nearby Skaigh had two in a garden on the moor edge on 28 February 1996, and birds were also seen that year at Walkhampton and Clearbrook. A pair were by the church at Harford in August 1997, and birds were seen in Widecombe in the early winter period of 1998. Also in this winter period, the first pair turned up at Thornworthy, although they departed in March without any breeding attempt. By 2001 birds were present in the summer at Thornworthy.

By the end of the 1990s, the population was still expanding into new areas, and some of these were rather isolated spots. On the 3 October 1999, one was found near buildings at Whiteworks. Another was at Prince Hall on 29 May 2000. Mary Tavy and Princetown had records that year, and by 2001 numbers were increasing rapidly in Princetown. A pair bred at Dartmeet in 2001 and again in 2002. Other records in 2001 came from Huccaby, Holne, Corndon, Lydford, Peter Tavy, Christow, and Gallant le Bower. Pairs were still finding new areas in 2002, or perhaps were being seen in areas where they had been missed before. One was at Beardown Lodge, Two Bridges on 7 September, and other sites not reported before were Jurston, Belstone, Ensworthy and Shapley.

A bird found dead at Okehampton on 24 June 1973, had been ringed in Belgium on 9 August 1971.

TURTLE DOVE *S. turtur*
Very scarce passage migrant. It may have bred on the eastern edge of Dartmoor in the past but proof is lacking.
D'Urban and Mathew found the species far from numerous in Devon, with most records from the east and north of the county. Today, Turtle Doves are almost a rarity, with the numbers in Britain reduced greatly. In Devon they still breed very sparsely in the east, but the nearest regular breeders to Dartmoor are the 10 or so pairs on the Haldon Ridge.

Most Dartmoor records are linked to the spring migration. There have been about 30 records since 1930, involving at least 38 birds. Three of the records have been in late April, with one at Burrator on 19 April 1980 being the earliest. Nineteen of the remaining records, involving 31 birds, have been between mid-May and mid-June. There have been two records in July, probably involving wandering non-breeders. A bird was flushed at Runnage on 5 July 1958, and two were at Bellever on 12 July 1957. In addition, the singing of a bird in July 1967 at Fernworthy presumably was an indication of a lonesome non-breeder, rather than of an occupied breeding territory.

Most of the records were in the years up to 1970; there has been a mere handful since. Sightings have usually involved singles, but four were at Hennock on 4 June 1951, and three were at Okehampton on 10 June 1931, and Sampford Spiney on 14 to 16 May 1979.

RING-NECKED PARAKEET *Psittacula krameri*
Very rare visitor.
On 20 February 2004, a bird was watched flying around the trees and fields in the Walkham valley, by Merrivale Bridge. This was apparently the first and only Dartmoor record.

CUCKOO *Cuculus canorus*
Migrant breeder and passage migrant.

On Dartmoor, the first arrivals each year are usually between the 17 and 24 April. The earliest record is of a bird on the West Okement at Black Tor on 1 April 1994.

This is a species that, over the centuries, people took for granted. It arrived at a certain time, was vociferous for a period, went quiet, and then departed. It was common and known by all, and up until quite recent times no great attention was paid to its numbers. Survey work for the *Devon Tetrad Atlas* between 1977 and 1985 showed that the species was unrecorded over large upland areas of Dartmoor (Sitters 1988). The *Dartmoor Moorland Breeding Bird Survey* of 2000 showed much the same picture, with birds absent from large areas of high blanket bog. The overall population in 2000 was estimated to be approximately 115 +/- 72 singing males. Birds were sparsely distributed, but were particularly associated with grassland and fragmented heath which contained valley mire. The population was smaller than expected, given the very healthy population of its main host species the Meadow Pipit (Geary 2000).

Casual recording over the years has shown many favoured areas. The valleys surrounding Burrator Reservoir have always proved attractive. The large stone outcrop in the Deancombe valley has always been known as Cuckoo Rock, and confirms the common status of the species in the area. Between four and six males have been heard in the area on many May mornings. No area supports more Cuckoos than the deep heather-lined valleys and gullies around Warren House Inn, Vitifer and Headland Warren, the habitat here proving just right for several host species. Several of the highest counts for the moor have occurred here. Ten were present on the 4 May 1980, up to seven in May 1989 and 1991, and ten again in May 2003. In 1990 10 to 15 calling males were present in spring in the study area around Postbridge, that partly included this area (Dare 1996). Other high counts have included up to

Cuckoo

nine in the Meldon valley in June 1970. At smaller sites, the Harford Common Bird Census has recorded two territories regularly between 1990 and 2000, with three in 1989 and four in 1999. In a tetrad square at Bonehill near Widecombe, four males/territories were located on 12 May 2002. To sum up, the breeding population on Dartmoor seems to be good in parts. Although numbers vary from year to year, birds can be regular, at times almost common, in certain favoured areas. They are probably not uncommon in a range of other moor edge and valley habitats, but are almost absent from other areas despite an abundance of host species. As well as Meadow Pipit, other host species noted have included Skylark, Wren and, surprisingly, Yellowhammer.

During late June and July, the bulk of the adult population departs quietly for the south. Juveniles are about for another month or so, before they too depart. Particularly late birds were at Cosdon Beacon on 8 September 1995, Yelverton on 12 September 1985, and the latest ever at Rippon Tor on 1 October 1997. Most departing juveniles are seen singly, but two amazing instances of flocks have been recorded. At South Brent on 17 August 1974, a flock of about 30 was seen in a *Macrocarpa* tree. They flew off northward and were followed by another ten. At the same place on 8 August 1979, a close flock of 30-50 birds were watched before they flew off westwards (Hurrell 1980).

BARN OWL *Tyto alba*
Very scarce resident.

The decline in the Barn Owl population of Britain, was first noted in the late nineteenth century. The reasons for this early decline are complex. Changes in agricultural practice, persecution by gamekeepers, and several severe winters during 1860 to 1900, all undoubtedly played a part.

This decline was noted by D'Urban and Mathew in Devon generally, but it is difficult to assess how the Dartmoor population fared. Because much of the high moor offered only unsuitable habitat, they must always have been patchy in their distribution. Where agricultural land spread out into the moor with adjoining farms, birds probably formed pockets of reasonably high density, but elsewhere they must have been very much restricted to moor edge habitat. In the early 1930s comments were made in the *Devon Bird Reports* that in moor edge parishes of Ashburton, Belstone, and Okehampton, Barn Owls were comparatively plentiful, and in Sticklepath it was even thought they were increasing. How representative this was of the moor edge Dartmoor land as a whole is unknown.

The severe winter of 1947 was disastrous for any pairs breeding on high ground. Prior to this winter, birds were resident on several farms in the Postbridge area, but none bred afterwards, and the species remained only a scarce visitor until 1957. A barn at Runnage was occupied in 1958, and breeding was proved there in 1962 and 1966. There were more frequent sightings in the area after 1961, including a pair at Broadaford in 1961 and 1965, and a wintering bird at Challacombe in the 1964-1965 and 1965-66 winters (Dare and Hamilton 1968). Although reports were non-existent, it must be assumed that most of Dartmoor's population suffered in a similar way at this time.

Between 1961 and 1964, the Devon Bird Watching and Preservation Society conducted an Owl Enquiry. As would be expected, most of the Enquiry's findings dealt with birds from lowland Devon, but there was some interesting information on Dartmoor's small population. The pair at Runnage was studied in some detail, and pellets analysed, giving an insight into the prey of the upland owls. Pellets collected in 1958, 1962 and 1963 contained a high percentage of Short-tailed Vole remains. The percentage was between 31.8% and 47%. Remains of Shrews were also much in evidence, with percentages of between 36.3% and 62.6%. Wood Mice remains were few considering the closeness of the site to conifer plantations, and no House Mice remains were found. Small birds and frogs apparently made up a very small part of the birds diet. Analysis of pellets from Lower Longford Farm, Moorshop took place in 1963, and there again Shrews were popular, forming 46.7% of remains. Short-tailed Vole formed 37.2%, Wood Mice 8%, and Bank Vole 7.3%. The Enquiry concentrated many minds on owls, and consequently some unexpectedly good areas were found. In the Dartmoor region this included five occupied farms around Moretonhampstead. The outcome of the Enquiry, as regards Barn Owl, was that although it was nowhere common in Devon, searching proved it commoner than expected (Goodfellow 1961-1964).

During the *Tetrad Atlas* years of 1977 to 1985, the fieldwork revealed a further collapse of the population after the hard winter of 1978/79. Around the moor, four squares were occupied around the northern edge from Sourton to Sticklepath, a pair bred in the Dunnabridge area, and around Fernworthy, and a probable pair was noted at Cornwood.

As numbers fell, efforts were made to boost the population with captively bred birds. This then tended to cloud the picture somewhat, and for many years it has been almost impossible to know if a sight record of a bird in a new area refers to a truly wild, or a captive-bred released bird. Barn Owls are very vulnerable to traffic as they glide low over roads, and unfortunately since the late 1980s there has been an increased number of road victims. What proportion of these are released birds is not known. Several, though, were ringed as young in the nest, and although the road deaths of these birds are to be greatly deplored, the recovery of certain ringed individuals has helped with our knowledge of Barn Owl movements. Birds are thought to be largely sedentary, but some have been shown to wander widely. A bird ringed as young at Okehampton in 1982 was found dead on the road in 1983 near Exeter, 34 kilometers away. Another ringed at Ashburton in 1987 was found on a road near Plymouth in 1988 30 kilometers away. On a happier note, a young bird ringed at Ponsworthy in 1993 was retrapped at a nest site near Bicton 39 kilometers away in 1995. Most interesting of all was one ringed at Cornwood in 1991, as a young bird in the nest, and then retrapped as a breeding adult near North Tawton in 1997. Not only had it wandered 43 kilometers, but it had travelled right around, or over Dartmoor, south to north!

In recent years, birds have been seen again in areas where they were reported 40 or 50 years previously. Although numbers are still low around Dartmoor, and breeding pairs very scarce, there is perhaps just room for a little optimism. The West Dart area now appears to be occupied again, with birds breeding at Bachelors Hall in 2001, and being seen nearby at Parsons Cot, Prince Hall and Princetown. A pair were watched at Pizwell, close to the old Runnage site, in 1999, and again in 2000. Three sites were found around Okehampton in 2001, and others have been seen in the area at Sourton Down and Bridestowe. The old site on the western slopes of the moor at Pork Hill and Moorshop, is producing records, and one wonders if birds have not always been present there. Records of hunting birds have also come from the Walkham valley at Routrundle, Babeny and Hexworthy.

Let us hope that this lovely bird is slowly regaining some of its former status around the moor.

SNOWY OWL *Nyctea scandiaca*
Very rare vagrant.
D'Urban and Mathew tell of a male that appeared at Ditsworthy Warren in the Plym valley on 13 March 1876. The warrener's eight-year-old son, having seen the bird take a rabbit, shot it as it passed overhead. It was sent to Mr. Gatcombe, the taxidermist in Plymouth, who confirmed it to be a male, by its size and overall white plumage. He also thought it to be an old bird. Others were reported in the press to be on Dartmoor at this time, but the only other confirmed bird in the general area was a large female trapped on Exmoor on 22 March (Ballance and Gibbs 2003).

LITTLE OWL *Athene noctua*
Very scarce resident.
Up until the 1840s Little Owls were occasional visitors to this country from the continent. From that time there were many attempts at introduction, with birds being released in several counties, mostly to the east and north (Holloway 1996). The early releases were not that successful, but later releases proved different, and the species gradually became established, and spread westwards.

At the end of the nineteenth century D'Urban and Mathew were uncertain whether any of the Devon records referred to truly wild birds. Some were introduced through release schemes, others were sold at markets and then kept in captivity. These were particularly adept at catching Cockroaches. Escaped birds could thus add to the confusion over their true status. The earliest record for the Dartmoor area would appear to be one near Ashburton in 1809, and this may have been a wild visitor from mainland Europe. However, it was still very much a rarity in the early twentieth century, and around the moor few were seen before about 1925 (Hurrell 1947).

By the early 1930s there were signs of expansion, and the *Devon Bird Report* for 1933/34 stated that numbers were increasing around the edge of the moor, and that birds were frequently being caught by trappers and farmers when ferreting the rabbit holes that they were thought to mainly inhabit. The increase continued into the 1940s, but pairs that had ventured out onto the moor were severely hit by the 1947 winter. Birds had regularly bred in the rabbit warrens around Postbridge, but none were seen after this bitter winter. It was 1958 before birds were again seen in the area, and two of these were found dying in snow in the West Webburn valley during January/February. Another was seen hunting at Runnage on 23 April 1958, but since then there have been no records in the area (Dare and Hamilton 1968). The savage 1962/63 winter must also have taken a heavy toll, and in the 1961/64 Owl Enquiry organised by the Devon Bird Watching and Preservation Society, their status in the county had declined to that of a very local bird, nowhere common. Birds were noted at certain places around the moor edge though, and the area around Chagford was thought to be particularly well-blessed (Goodfellow 1961-1964).

By its very nature this species is not always easy to locate, and in areas like Dartmoor, where pairs are often widely separated, there may be little calling or territorial activity. In these situations pairs are very easy to miss. The Devon *Tetrad Atlas* showed a grouping of records around the southwest edge of the moor (Sitters 1988). In 1981 a concerted effort in the Yelverton to Cadover Bridge area found birds at five sites. Twenty years on there may not now be that number of pairs in the area, but the southwest edge still produces records, with well-watched birds at Criptor Newtake, and below Cox Tor, with another territory close by, at Wedlake Farm. Recent records from the rest of Dartmoor are rather sparse. Birds are still seen now and again by Yardworthy Farm, near Fernworthy, and farmers reported in 1996 that there were three pairs in the area (R. Waller *pers.comm.*1997). A traditional site near Halshanger Cross had a record in February 1999, and birds have been seen at Ashburton, Okehampton and Sticklepath. Although it is now without doubt a very scarce bird, it is almost certain that there are a few established territories around the moor edge that have yet to be found.

TAWNY OWL *Strix aluco*
Resident breeder.
D'Urban and Mathew knew them as common in the woods and plantations of Devon, but thought they were not as plentiful as they had formerly been. No specific references were made to Dartmoor.

They are at present a widespread, but greatly under-recorded species. On Dartmoor they inhabit most types of woodland, and also agricultural land with sufficient hedgerow tree cover. It would appear to be probable that since the late twentieth century, when many of the conifer plantations matured, numbers increased on the moor. Due to lack of records over anything like a representative area, this is impossible to confirm. It is thought likely, however, in the Postbridge area, where the population could now be higher than the 15 to 20 pairs present from 1956 to 1967 (Dare 1996.)

Elsewhere there have been few estimates of numbers, but the details that have been noted would seem to indicate a healthy population.

None of the large plantation areas have been properly surveyed, but some indications have been gained in recent years by groups out at dusk looking for Nightjars. At Fernworthy up to six were located on 18 July 1997, and three or four territories were found in the Assycombe Valley area alone in July 1999. Up to three territories were found at Bellever Plantation in 1998 and 1999, and at least two territories were noted in Soussons Plantation in 1998. Birds were located in four areas around Burrator Reservoir in 1997 and 1999. At Yarner Wood five males were present in 1987. Three or four pairs bred in 1988 and 1989, but then only two in 1990, and this number had not increased by 1996. At least two pairs bred in Hembury Woods in 1999 and 2000, and in the agricultural Common Bird Census plot at Harford, one territory was located in 1990 and 1991.

For nest sites, birds usually use a suitable tree hole, but nest boxes have been accepted at Yarner Wood and Postbridge. Old nests of Buzzard, Crow and Magpie were used in the Postbridge area in the 1950s and 60s, and once birds nested in a barn (Dare 1996).

LONG-EARED OWL *Asio otus*

Very rare passage migrant and winter visitor. Has probably bred in the past, and may still do so occasionally.

In about 1958, a young Mike Sampson went on a picnic to Skaigh, near Belstone. Whilst exploring the area he came across a tree with a nest in it, about 10 feet from the ground. Sitting in the nest was a Long-eared Owl, with ears fully erect. Although the outcome of this nesting attempt is not known, this appears to be the only known nesting attempt on Dartmoor.

With a species that is quiet for much of the year and almost totally nocturnal in its habits, it is more than a possibility that the odd breeding pair may have been overlooked. The plantations and natural woodland around the moor would appear to offer ample suitable habitat, but pairs breeding at a very low density would be very easy to miss. There was an increase in breeding numbers in Britain generally in the latter nineteenth century, that was followed by a decline in the twentieth century (Holloway 1996). One of the reasons for the decline may have been competition from the larger Tawny Owl, and this could be one of the reasons for the bird's rarity on Dartmoor, where every plantation, and most other woodland, holds a healthy population of Tawnies.

D'Urban and Mathew knew the species in Devon only as an autumn migrant and winter visitor, but did mention a pair being shot at Buckland Abbey in April 1846. The first recorded bird on Dartmoor proper in the twentieth century was one seen low down and pressed against the trunk of a larch in the Haytor area on 1 August 1942. Another bird, in a similar upright stance, was seen in a conifer in Brisworthy Plantation by the writer's grandfather in the late 1940s. After this, none were reported until a road casualty was found near Ashburton in February or March

1979. One was disturbed from thick cover at Fernworthy Reservoir, on 7 November 1990, and another was found roosting in an overgrown hedge at Tottiford Reservoir on 1 February 1992.

SHORT-EARED OWL *A. flammeus*
Very scarce passage migrant and winter visitor.

The Short-eared Owl has always been scarce on Dartmoor. During their ten years Snipe shooting on the north and east moor, D'Urban and Mathew never came across a bird. They knew of no nesting attempts, and although the young plantation areas in the early to mid-twentieth century could have provided suitable habitat, breeding has never been suspected. Perhaps the Field Vole population, its main source of prey, has never been sufficiently high or stable.

After the breeding season, the first early arrivals have probably been dispersing juveniles from natal areas elsewhere in Britain. The earliest was a bird at Rippon Tor on 25 August 1990. Other early records were of one at Brockhill Ford on 5 September 1957, and one at Zeal Down on 12 September 1978. The first main wave of migrants from the continent shows in Dartmoor records between the dates of 21 October and 14 November. They usually appear singly, and may well be birds moving westward through the country after crossing the North Sea and making landfall on the east coast. They have been seen in many areas with sufficient cover to hold prey, but the Birch Tor/Vitifer/Warren House Inn area has always proved particularly attractive.

Most have been one day sightings, suggesting that the birds were moving through the area. Other records have occurred over the years between the dates of 26 November and 6 December. By then birds are seeking wintering habitat. They may well be in the area for the rest of the winter, but as they have large hunting territories, may only be seen intermittently, at best. Wintering birds can be encountered hunting low over marshes and moorland, throughout January and February, although records are not annual. There is a distinct increase in birds appearing in March, and this no doubt reflects an early movement back towards the breeding areas. Definite migrants have been seen on several occasions in the last two weeks of April, and May records are not unknown. A bird stayed in the Postbridge area from mid-April through to 1 May 1979, and that same year birds were still hunting at Muddilake into May. The latest May record was of a bird near Postbridge on 18 May 2002. In 1986 a bird was watched in suitable breeding habitat at Wollarford Down on the unprecedented late date of 4 June. It was not seen subsequently, however.

There have been two marked influxes of birds in the last 30 years, both no doubt linked to exceptionally good numbers of Field Voles being present. The first occurred in late October 1978. Birds were noted far more regularly than usual over several areas of the moor, and it soon became evident that an unusually large number of individuals were present. The main area of sightings were the bogs and marshes of the central basin. Birds were regular over Muddilake, the Cherry Brook, Two Bridges, Powder Mills, and south to Princetown, the marshes around Black Tor and Foxtor Mires. Individuals hunted over large areas, but as winter progressed it was apparent that there was rivalry for areas, and suggestions of birds having territories. Mid-air clashes between hunting birds, appearing from different directions, were noted more than once. Other birds were seen regularly around Riddon Ridge, Cator Common and Bellever during the 1978/79 winter, and also at Rippon Tor, Corndon and Yar Tors, and Cornwood. The main area around the central basin held at least eight different birds, and the total could have been higher. With the other areas included, it is likely that the wintering number on Dartmoor that winter was in the range of 12 to 15 birds. Most appeared to have moved off by the end of March 1979, but one or two hung on into early May.

The second influx was noted over a shorter period. In January 1999 up to four birds were seen in the Rippon Tor/Halshanger Common area. Almost daily watching here over the next two months or so proved that at least double that number were in the area. Birds appeared either hunting over the *molinia* covered bog area of Halshanger Marsh or the southern slopes of Rippon Tor in the late afternoons and had usually disappeared back into cover by last light. To get a total present it was necessary to get all the birds in the air at once, and this of course proved impossible. Eight were seen in the air together on 9 March, and it is likely that the total was about 10 to 12. Numbers declined during late March, and the last birds were seen on 13 April. This occurrence was different to the 1978 influx in two respects. Firstly it was a mid-winter gathering, with no sign of any out of the ordinary numbers the previous autumn, and secondly the high numbers were largely all at one site. Other birds, which may well have been additional, were seen at Skir Hill, Merrivale and Ryders Hill, all in

Short-eared owl

February. One or two birds seen at Statts Bridge and Powder Mills on 22 to 26 April that year, were probably migrants, although they could have been the last of the Halshanger birds moving off.

NIGHTJAR *Caprimulgus europaeus*
Scarce migrant breeder and passage migrant.

D'Urban and Mathew remarked that Nightjars were common about woods, ferny heaths, and orchards, especially on the borders of Dartmoor. In the 1920s churring males were noted in the area around Thornworthy, where they nested in the good growth of heather present there at the time (R. Waller *pers. comm.* 1997) At the same time they were also being seen each year in the Deancombe valley, Burrator (the late H.G. Hurrell *pers. comm.* 1920) In the 1933 *Devon Bird Report,* an observer stated that they were rarely found on north Dartmoor, except in warm-lying ferny coombs.

Nationally many conifer plantations were brought into being in the years after the First World War, and these young plantings were found to be very acceptable to Nightjars. However, by the 1930s these areas were becoming less and less attractive for nesting. As the trees matured, the canopy closed, and birds were unable to get to the bare earth nesting sites below. At the same time traditional habitat was being lost to agriculture and urbanisation, and a decline in breeding numbers began.

On Dartmoor, the few records from the 1930s and 1940s reflected this decline. Birds were still using the 'ferny coombs', but there were no plantation records, although parts of Fernworthy may well have been occupied. During the early 1950s birds were being seen or heard quite regularly around Haytor, and in 1954 a pair bred at Yarner Wood (Page 1992). They were recorded there until at least 1958. Also in 1954, a male was heard churring at Wrangaton on 30 June, the first for many years. Other occupied areas at this time were Buckfast, Hembury Fort, Shaugh Prior and Newbridge, Holne.

What was probably Dartmoor's first plantation record occurred in 1957 when a male was churring in Bellever Plantation on 19 June. Birds were then annual here until 1960, and possibly until 1965 (Dare and Hamilton 1968). Amongst the young trees in nearby Soussons Plantation, birds were first heard in 1958, and records from here continued until 1965, when five were present in July. After this year there were no records from either plantation, presumably because maturing trees made the habitat unsuitable. In 1961 and again in 1966 males were heard close by in the mine gullies around Vitifer. Reports came from Lydford Forest in 1969 and 1971, and from Sticklepath in 1972. A male was heard at Yarner Wood in June 1972, apparently the first since 1958. In 1976 Burrator was being occupied, and although four were present in 1978, only one could be located in 1979, and none thereafter.

During the 1980s the decline in the British breeding population stabilised, and numbers then slowly began to increase. At Yarner Wood birds were again present from 1988, and at the adjacent Trendlebere Down one or two pairs were annual from 1982 to the present time. Three pairs were here in 1994 and 2002. A bird was churring in the unusual locality of Sampford Spiney in May and June 1980, but the real increases on the moor occurred in the plantation areas that had recently been clear-felled or newly replanted. Throughout Britain, the harvesting of plantation timber gave a great boost to the Nightjar population. Huge areas around the country again became suitable for nesting birds.

The most dramatic increases on Dartmoor were in the following plantation areas:

Fernworthy: First noted here in August 1981, numbers were thought to be two or three pairs throughout the 1980s. By 1992 five pairs were counted, but it was 1997 when members of the Dartmoor Study Group mounted a survey of the whole area, that the full picture emerged. In June of 1997 23 churring males were found. Annual counts since then have produced totals of between 12 and 28. Counts have never taken place before mid-June, to obviate the problem of any migrant males still being in the area, but even then counts have always been subject to weather conditions. Some of the lower counts have been in less than perfect conditions, and are probably not representative of the number of males present. Churring males do not always equate to successful breeding, as in 2000 when 15 males were counted, but only eight pairs were later known to have young.

Bellever: Eight males were located by the Dartmoor Study Group in June/July 1998. Further yearly counts have shown between six and 11 present, although this number had increased to 20 in 2004. The same caveat applies regarding weather conditions, and it was noted that although 10 males were found in 2000, only six pairs later produced young.

Soussons: The Dartmoor Study Group found four churring birds here in 1998, then an increase to between five and nine each year to 2003. This number had risen to between 16 and 18 in 2004, an exceptionally good year for Nightjars.

Night jar

Other significant increases have been noted at Hennock, where eight males were found in a recently clear-felled area in 1992, with five in the same general area in 1994, Burrator, where at least four males were present in 2002, and at Roborough Down, where at least three or four have been seen each year since 1999. Other areas that have had one or two pairs recently have included Aish Tor, Wooston Castle (Drewsteignton), Ausewell and Padley Common (Chagford).

It would look probable that the average total number of churring males holding territory at present on Dartmoor would be about 50/60 birds, although the BTO Nightjar Survey of 2004 produced a higher total of 72. However, as the past has shown plantation habitat is transient, and in maybe five to 10 years as the present new plantings get bigger and conditions change, the population may again decline.

Birds tend to arrive and leave unseen, but migrants out of their usual habitat have included singles at Rattlebrook Hill on 18 August 1951, Ugborough Beacon on 9 September 1991, Dendles on 6 September 2001, over the A38 near Ashburton on 16 August 2002, and hawking insects over the road at Pixies Holt on 16 September 2002.

SWIFT *Apus apus*
Migrant breeder and passage migrant.

The first Swifts usually arrive over Dartmoor in the last few days of April or during the first week of May. The earliest recorded have been single birds at Buckfastleigh on 1 April 2000, and Sticklepath on 7 April of the same year. In 1962 a very interesting and early record concerned two parties of 50 to 60 seen at Devils Elbow and near Postbridge on 8 April flying northwest into a strong wind. They would occasionally drop and circle close to the ground giving good views.

Swifts are almost totally reliant on buildings for nest sites. Therefore, not surprisingly, most are found in the towns and villages around the moor. They do not appear to be put off by altitude, and several pairs have always nested around Princetown. In 1989 they were noted in Princetown as abundant, with colonies exceptionally strong. Since then there has been something of a decline with eight to 10 nests in 1992, and less since. As elsewhere, this is probably due to the birds being denied traditional nesting sites by building alterations and general renovation work.

During the summer, especially in late June, large congregations will assemble to feed on flying insects. These feeding flocks are usually associated with low cloud and thundery conditions. On 21 June 1986 700 or more passed north over Okehampton in fifteen minutes, and the next day 200 were over Bridford. Several hundreds appeared over Horrabridge on 25 July 1999 feeding on flying ants, and 250 were noted there on 21 July 1961, no doubt taking advantage of the same food source. On 20 June 1960 about 200 were feeding over High Willhays.

Although buildings are used almost without exception for nest sites, there have been instances of birds showing interest in more unusual sites. D'Urban and Mathew remarked that they had been informed of Swifts nesting in stone boundary walls on

Dartmoor in the 1870s. They made little comment, and it is difficult to know how much belief to have in this information. A more interesting and far more probable site involved birds at Merrivale Quarry. On 15 July 1981 one flew into a transverse fissure about two feet long and two inches high, then after a few minutes went off south down the Walkham Valley. After 25 minutes a bird returned, entered the crevice, and remained four minutes before departing. The observer had no doubt that young were being fed.

Birds usually leave in late July and early August, with most gone by mid-month. Late stragglers are seen until the end of August, and September migrants have been noted at Harford Moor on 15 1991, Okehampton on the 18 1982, and at Haytor the same day.

KINGFISHER *Alcedo atthis*
Very scarce resident breeder.
The Kingfisher is the quintessential bird of pools, lakes and slow flowing rivers, and as such, suitable habitat for it on Dartmoor is very restricted. The actual number of pairs breeding within the area of the moor is unknown, but is undoubtedly very small. When breeding has taken place or been suspected, it has always been on a reservoir site, or on the lower, quieter, reaches of a river. Of the reservoirs, Burrator has had most records, with pairs suspected of breeding in 1981 and 2002, and birds seen most years. Fernworthy had birds in the breeding season in 1990, but not since. On the rivers, breeding is regular on the Walkham between Horrabridge and Double Waters, and in 1985 two pairs bred on the river near Grenofen. On the Dart, birds are seen often around Buckfastleigh, and breeding was confirmed in 1999 at Hembury Woods, with a pair back at the same site in 2000, and a pair probably breeding at Spitchwick in 1992. Breeding or probable breeding has also taken place on the Plym at Shaugh Prior in 1986, on the Teign at Steps Bridge in 1992, on the Meavy with one territory on the Water Bird Survey site in 1996, and on the Bovey, with a pair rearing two broods at Parke in 1999. During the survey work for the *Devon Tetrad Atlas* 1977-85, a pair was confirmed breeding on the Erme, above Ivybridge.

Most sightings of birds on the moor refer to the period directly after the breeding season when the fledged young disperse. They then follow water courses high up onto the moor, and account for sightings from July to December, on stretches where they do not occur at other times of the year. A good example of this is on the West Dart, where birds have been seen quite regularly at Huccaby, and to a lesser extent Dunnabridge, Prince Hall and Two Bridges. There was also a record of a bird as high up the river as Wistmans Wood in late August 1997. Even small streams are followed on occasions, as with the birds on the Glazebrook on 20 June 1942, and the Narrator Brook at Outcombe Wood on 11 July 2000.

Individuals will also return again and again to a favoured spot if it is a good source of food. Goldfish ponds were raided near Whitchurch Down in 1961 and at Grenofen in 1993, and a small pond on Roborough Down near Yelverton was visited for aquatic insects and possibly newts in 1929/30. Two farm ponds in the Widecombe area were visited from 1996 to 1998. At Littlecote Farm a bird was first seen in November 1996. It came to a pond almost daily for an hour or more, and caught mainly newts. It was seen until 9 January 1997. It, or another, returned on 14 January 1998, when it stayed for six days. At Broadapark Farm, a bird, perhaps the same, was seen quite regularly at pools during 1998.

BEE-EATER *Merops apiaster*
Very rare vagrant.
D'Urban and Mathew note two that may, or may not have just come within the Dartmoor area. A bird was seen at Ivybridge in 1822, and another was shot in the Ashburton area. The only record in recent times was of a bird seen on 23 May 2002. It was watched on wires early in the morning at Brisworthy Farm, near Cadover Bridge. It was studied for about 15 minutes, during which time the observer saw the bird eject a pellet. Unfortunately it was not about later in the day.

ROLLER *Coracias garrulus*
Very rare vagrant.
The inclusion of this vagrant from the south as a Dartmoor bird is down to one record of a bird found near Chagford on 3 August 1911 (Moore 1969).
It was present for several days.

HOOPOE *Upupa epops*
Very scarce vagrant.
Every year in the spring Hoopoes, returning from the south of Europe, over-shoot the French coast, cross the Channel, and find themselves on the south coast of Britain. Most birds are seen close to the coast, but in the last 70 or so years about 50 have been reported from Dartmoor.

The earliest arrivals in Britain are usually in the latter half of March, and Dartmoor has had four at this time. Birds were found on 14 March 1957 at Horrabridge, and on the same date in 1981 at Sampford Spiney. Another was at Yelverton on 23 March 1960, and a fourth at Bittaford on 26 March 1996. The main arrival, subject to weather conditions, is about a month later in late April, with some records in May. Twenty one Dartmoor records have been in this period, and this included the largest multiple sighting, when four were around Bellever Common at the end of April 1964.

Breeding has never been suspected, so the birds seen occasionally in June and July were probably individuals that had arrived earlier in the spring and had not found their way back to continental Europe. Dartmoor has seen only a few autumn records, and the latest records in October have involved one at Liverton on 13 October 1982, one at Ivybridge on 17 October 1988, and a long staying bird at Fernworthy from 13 to the 22 October 1995.

Hoopoes can appear almost anywhere where there is short grassland or lawns for them to feed. On Dartmoor most of the records have been around the southern and southwestern fringes of the moor, although most areas have had records. One area with an unexplained high total of records has been the area of Horrabridge, Sampford Spiney and Grenofen, where over the years at least 12 of Dartmoor's sightings have occurred.

WRYNECK *Jynx torquilla*
Very rare passage migrant. Has bred.
The Wryneck, once a widespread and fairly numerous migrant breeder in much of Britain, declined to the point of virtual extinction during the twentieth century. It was never as regular a breeder in the southwest as elsewhere in southern England, so we are perhaps fortunate to be able to claim two breeding records for the edge of Dartmoor. In 1948 a pair was recorded breeding at Crapstone near Yelverton. They were reliably reported by F. Howard Lancum, who was a resident of Crapstone, and a reporter of many interesting bird records in the early twentieth century. He noted a bird again in the area between 26 May and 2 June 1954, but breeding was not then suspected. Another instance of breeding in the same general area was in 1949, when a pair bred on the edge of West Down (Harvey Kendall *pers. comm.* 2004*)*. The next, and only other instance of a male taking up territory was in 1986, when a bird was present in a small area of oak wood on the edge of Lee Moor village, calling persistently, between 5 and 22 June.

In total there have been about 15 records from Dartmoor during the last 70 or so years. Excluding the breeding records, there has only been one spring migrant, and that was a bird at Hennock Reservoir on the early date of 6 April 1987. An early autumn migrant was at Wrangaton on 1 August 1965, and the latest autumn bird was an individual in an oak wood near Deancombe, Burrator on 18 October 1975. The bulk of the autumn records have occurred between 25 August and 20 September, a classic migration period for this species in the Southwest. At least four of the autumn records have been from rural gardens, where no doubt ants were the main attraction.

GREEN WOODPECKER *Picus viridis*
Resident breeder.
This species is widely distributed in all woodland habitat around the edge of the moor, and at suitable areas in the interior. There appears to have been a slow but constant increase over the years, and Dartmoor probably holds the best breeding numbers in the county. An indicator of their relative abundance was the figure reached in 2000, when the Dartmoor Moorland Breeding Bird Survey, coupled with casual observations from members of the Dartmoor Study Group, produced a total of at least 53 pairs/territories from 44 areas.

In the Postbridge area, three to five pairs bred from 1956 to 1962, but were wiped out in the severe 1962/63 winter. Birds were slow to return and it was after 1967 before the previous total was restored. By the mid-1990s, breeding pairs numbered four to six (Dare 1996). At Burrator, where the mixed woodland and plantations around the reservoir and in adjacent valleys, offer excellent habitat, Green Woodpeckers have always been numerous. Four to six areas have been occupied most years since the early 1990s, and a possible 10 territories were occupied in 1996. Two pairs have been found for several years at Hennock Reservoirs, with three in 2000. The Harford Common Bird Census site held two pairs regularly throughout the 1990s, but a third territory was recorded in 2000.

Birds will on occasions wander well outside their woodland territories to feed. A bird at Aune Head on 30 July 1942, and one at Cranmere Pool on 26 August 1963, were extreme cases of this feeding activity.

Great spotted woodpecker

GREAT SPOTTED WOODPECKER *Dendrocopos major*
Resident breeder.
D'Urban and Mathew and other authors of the nineteenth century all stress the scarce status of this species in Devon. Birds were apparently met with more in winter and spring than at other times, and it is stated that with certainty it was the scarcest of the three English woodpeckers in the West Country.

Things have certainly changed since then. During the twentieth century numbers increased greatly on Dartmoor, as elsewhere in the county, and birds became a regular sight. All suitable valley woodland was colonised, birds were noted in the ancient relict oak woods of Black-a-tor Copse and Piles Copse, although probably not Wistmans Wood, and conifer plantations attracted breeding pairs. The first record in conifer plantations would appear to be the bird seen in Baggator Plantation on 18 April 1937. Although striking in appearance, birds can easily be missed in the upper branches of woodland trees, and as no species survey has ever been carried out to establish numbers, the total for the Dartmoor area is not known. In 1999 the Dartmoor Study Group found birds at 29 sites, although due to uneven coverage this must be a considerable under-estimate, and it is probable that about double that number of territories are present.

Areas that have been proved to hold a healthy population include the wooded valleys around Burrator that have had three to five pairs regularly since 1994, Yarner Wood which has had between two and six pairs regularly from at least 1987, with probably 12 pairs in 1990, and Hembury Woods where three or four pairs have been recorded since 1996. The plantations around Hennock Reservoirs have also had three or four pairs since 1992. At Fernworthy birds were found in three areas in 1999, and five areas in 2002. The Harford Common Bird Census site noted its first birds in 1988, and there have been one or two territories ever since, with 1989 producing three. At Dunsford Wood Nature Reserve four pairs were located in 2002, an increase on the usual two or three. Survey work for the Devon Atlas Project in 2002 showed 12 pairs to be present on the south side of the Dart valley, between Venford Reservoir and Holne Bridge, including five in Holne Chase.

In the Postbridge area two or three pairs were regular between 1956 and 1962. However the winter of 1962/63 wiped out these pairs, and birds were only seen twice between 1963 and 1968, the first being a sighting at Cator Common on 12 July 1964. By the early 1990s the species was showing the expansion noted elsewhere, and eight to 12 pairs were present (Dare 1996).

With the felling of much commercial plantation in very recent years, and the consequential loss of breeding habitat, it is possible that the years ahead could see some reduction in numbers, but at present this does not appear to be the case.

LESSER SPOTTED WOODPECKER *D. minor*
Very scarce resident breeder.

In the nineteenth century all writers agreed that the Lesser Spotted Woodpecker was not an infrequent bird in the woodlands of Devon. Although not specifically mentioned, no doubt this statement also involved the border woodlands of Dartmoor. As the twentieth century progressed, the status of the species changed and it soon became very scarce. It can be very easily overlooked, especially in the winter and even in the breeding season when pairs are at very low density, so there could well be areas holding breeding pairs that are as yet unrecorded.

Unlike the other two resident woodpecker species, it is very much a bird of the lower decidous woods around the edge of the moor. It does not venture far into the interior, except along well wooded river valleys, and does not take to the conifer plantations. Burrator, so well liked by the larger woodpeckers, has only ever had three records, and the Postbridge area only one second-hand record when a bird was reported by foresters at Bellever in April or May of 1966 (Dare 1996). It has been noted in the past that the large proportion of Devon records occur in the rain shadow of Dartmoor, thus avoiding the wetter western areas (Sitters 1988). This is also true of the bird's distribution on the moor. Since the first *Devon Bird Report* in 1929 there have been records from some 46 areas. Although areas around the south and west edges of the moor are represented, the overwhelming proportion of records are from the valley areas of the Dart, Bovey and Teign on the eastern side. Locations of birds are given from the National Park boundary, upstream towards the river's source.

River Dart: Birds have been seen at Buckfastleigh in 1983, 1987, 1993 and 2001,whilst upstream at Hembury Woods birds were first recorded in 1984, although they had no doubt been in the area for much longer. Breeding pairs were surveyed annually from 1994 to the present, and a maximum of seven pairs were located in 1996. A breeding season record was obtained from woods at Holne Bridge in 1989, and Spitchwick had singles in 1978 and 1981, with a pair noted in 1980. Nearby at Leigh Tor a pair bred in 1982, and the area around New Bridge has had records in several years between 1982 and 2002. Birds were noted further up the Dart valley area in recent years with one at Bellpool Island in 2000, and singles in woods below Venford Reservoir in 1996 and 2000. At present the individual at Coombestone, Dartmeet in 2001 is the furthest upstream of any that have been seen. Slightly off the main river valley woods, but probably associated with them, were records of individuals at Scorriton in 1985, and at Gallant Le Bower, Holne in 2002. Also birds seen at Druids Wood, Ashburton in 2002/2003 probably originated from stock in the Dart valley.

River Bovey: Yarner Wood has always been the prime Dartmoor site for Lesser Spotted Woodpeckers. They were known to be present there before 1954 (Page 1992), but the first recorded breeding was in 1964. Pairs have been regular ever since, usually two or three, but at least five pairs in 1987 and four in 1994. Birds are seen occasionally nearby on the edge of Trendlebere Down, and in 1981 a male was seen in Pullabrook Wood. There was a record in 2000 from Houndtor Wood, and in 2001 a bird was seen at Woodash. Lustleigh has had records in 1948, 1979 and 1989, and others have been seen at Horsham Cleave in 1999, and Manaton village in 2002. Nearby on the Becka Brook a pair was watched in Leighon Wood in late April 1984.

River Teign: On the Teign there have been records from the National Park boundary at Dunsford to Dogmarsh Bridge and Chagford. Steps Bridge has had several records between 1978 and 2000, and a male was upstream at Clifford Bridge in 1981. Upperton Wood had a record in 1995, and Fingle Bridge has had birds on a regular basis every year or two since 1971 to the present. A pair were at Hannicombe Wood in 1984, and Drewsteignton claimed a record in 1981. In 1995 a male was seen at Dogmarsh Bridge, and birds were seen in the Chagford area in 1955, 1988 and 1994. Linked undoubtedly to the Teign population, were the birds that appeared at Fernworthy Reservoir in 1989. There have been records in most years since. At about 360 metres above sea level, these are unusually far into the moor, and the highest of the species on Dartmoor.

Away from the main river woodland sites, birds have been seen in several other localities. The wooded valley of the River Walkham has held birds at Grenofen in most years since 1981, with a record at Horrabridge in 1996, and downstream at Double Waters in 1984. The small woodland area around the home of the Hurrell family at Wrangaton has proved rather exceptional over the years. It has records spanning more decades than any other site on Dartmoor. The earliest dates are not known, but birds were certainly being seen in the 1930s. The bird or birds that were seen on four dates in 1947 were the first for several years, and the drumming that was heard in February 1957 was the first heard for twenty years. Birds continued to be seen up until at least 1992. Other records have come from Whitchurch Down in 1959, Ashburton in 1961, Belstone Cleave in 1964, Lee Moor in 1996, Hennock Reservoir in 1980, Goodameavy in 1984, Moorhaven in 1989, Christow in 1995 and New Waste, Cornwood in 2002. The three records at Burrator were in 1965,1970 and 1980.

Lesser Spotted Woodpeckers are residents, and move very little from their breeding territories, even in winter. Given this fact, and their very elusive nature, it must be at least probable that any one of the above mentioned records could represent a potential breeding pair, irrespective of the time of year seen.

WOODLARK *Lullula arborea*
At present a rare breeding resident, and passage migrant. Formerly much more widespread.
The Woodlark just manages to stay on the list of breeding birds due to its present stronghold in the Teign Valley coming partly within Dartmoor National Park Boundary. A few pairs breed around Bridford and the odd pair around Doddiscombesleigh could be just within the Park. In 2000 a pair raised two young near Widecombe, confirming that occasional breeding pairs can occur outside of their normal range, and may possibly go unrecorded.

In the past the status of the species was very different. It has always been a bird subject to surges in population, followed by quite rapid declines. In Devon generally, where the population is now woefully small, D'Urban and Mathew knew them as common residents in the late nineteenth century, and quoted Montagu, from the early nineteenth century, as saying that they were more common in Devon than anywhere else in England. A decrease was noted in the 1920s in Britain, followed by an increase in some western counties and an expansion of range in the 1940s. This expansion peaked in about 1951, and a decline was noted from about 1954. This decline accelerated from about 1960, and numbers were dealt a huge blow by the arctic winter of 1962/63 (Holloway 1996). Although the last comments referred to the British population as a whole, they equally apply to the numbers on Dartmoor.

They were always birds of the moor edge, with small woods and open heathy ground, rather than the central moor. Most of the records from the 1930s and 1940s period came from the southern and southwestern fringes of the moor. This could have been a reflection of genuine distribution, or the fact that these areas always had enthusiastic and competent observers. On 22 March 1931 six to eight birds were singing on the edge of Roborough Down above Grenofen, and a single singing male was located in the Walkham valley at King Tor the same day. The full stretch of Roborough Down, from the Walkham/Tavy confluence at Double Waters to the eastern edge overlooking the Meavy, proved very acceptable to Woodlarks in the 1930s and 40s, with observers noting them as well-established in the area. There were no records in the Burrator area in the 1920s or early 1930s (Smaldon 1982), but a singing male was heard in May 1937. Birds became regular here from the late 1940s, mainly on the slopes of Down Tor, in the Deancombe valley, and in the fields surrounding Lower Lowery Farm. In 1947 songflight was observed over Callisham Down, Meavy on 1 June, and a pair was found later the same month on nearby Wigford Down (L.W. Slade *pers. comm.* 1997) Further to the south, a pair nested at Wrangaton in 1938, and at Buckfast in 1947 they were recorded as being nearly as common as Skylark. On the eastern side of the moor during the late 1940s, birds were noted singing at night from the bracken covered slopes around Thornworthy (R.Waller *pers. comm.* 1997). As birds were widely distributed at that time, it is more than possible that many went unnoticed, or at least unrecorded.

In the 1950s there was a fall off in numbers. Birds were still being reported in the Burrator area, particularly in the Deancombe valley, where three singing males were heard between Snappers Tor and Combeshead in May 1958, but birds were no longer being reported from Roborough Down. Elsewhere, Yarner Wood had a singing male in late April 1954. What was undoubtedly a declining population was severely hit by the winter of 1962/63. Heavy snow and sub-zero temperatures for weeks on end dealt the species a blow from which it never recovered. At Burrator birds were still being seen, but in reduced numbers, and the last song was reported there on 24 May 1969. Reports during the 1960s were few and far between. Although birds were seen in some new areas, the population now appeared fragmented as well as greatly reduced. Singing males were present at Wrangaton in 1963 and 1966, Buckfastleigh in 1968, Belstone in 1965, and at Hennock in 1968. A bird was seen in the unusual locality of near Brat Tor on 1 July 1964.

In 1970 a small new, or previously unrecorded, population was found on the moor edge at Cornwood, especially around New Waste. Two or three singing males were noted there up until 1978, but by 1980 they could not be found. The area around Hennock Reservoirs had a pair or two from 1977 to 1989, and Chagford had singing males in 1970. 1981 and 1989. A bird sang at Meldon Reservoir in January 1981, and a pair bred at Hembury in 1989. The last record from the once well-populated area of Roborough Down was in 1975 when a bird was seen at Double Waters, although not too far away a bird sang at Sampford Spiney in June 1986. By the late 1980s what few records there were probably represented wandering birds, rather than established pairs, and by the 1990s the last remnants of population were to be found in the Teign valley, where the few pairs still exist.

The birds seen in central Dartmoor have probably all been wandering males that have taken up temporary territories and then moved on. The first was a bird at Brimpts Wood on 30 May 1937, whilst others have included nine records from the West Webburn valley and Cator Common areas between 1956 and 1962, and a bird on the West Webburn in May and June 1968 (Dare 1996). Breeding has never been suspected.

Recently there have been two records of migrants over-flying the moor in autumn. In 1998 two flew south calling over Grippers Hill on 18 October, and in 2001 a bird flew south over Huccaby on 17 October. Presumably these were birds from one of the migratory populations in eastern England, rather than from our own small sedentary population.

SKYLARK *Alauda arvensis*
Resident breeder, and passage migrant. Altitudinal migrant with the breeding population moving off the high moor in the autumn.

Recent surveys have shown that the Skylark is the second-most numerous breeding bird on Dartmoor, beaten only by the Meadow Pipit. It nests in varying numbers in all types of open grassland habitat, heather moor and blanket mire. As is usual with most species of common bird, there is very little information available in past literature regarding breeding numbers. This being the case, the professional surveys undertaken on moorland birds in the last 25 years are especially important and provide a yardstick for any future assessments.

In the last 30 years or so the Skylark population in this country has gone into rapid decline. As much as 52% of breeding pairs may have been lost in that period. This decline is largely linked with farming methods and loss of habitat on low agricultural ground. Populations on the highlands have not been so badly affected.

The survey in 1979 for the RSPB and DNPA showed that numbers on Dartmoor were very good. Birds were recorded in 86.8% of the survey squares, and the total population was estimated to be about 15000 pairs. The average density was 32.8 pairs per square kilometer. This density was noted to be much higher than similar upland areas surveyed in Scotland and Wales (Mudge *et al* 1981). An Environmental Baseline Survey for the RSPB in 1992 found 1515 individuals in 52 square kilometers of high northern Dartmoor, giving a density of 29 birds per square kilometer (Chown et al 1992). The DNPA High Moorland Sample Survey of 1997 located 512 birds in a 16 square kilometer area that had previously been surveyed as part of the 1992 survey area. It yielded a density of 32 birds per square kilometer. The same area during the RSPB survey in 1992 had 463 birds in a 15 square kilometer area, giving 30.9 birds per square kilometer density. The Dartmoor Moorland Breeding Bird Survey of 2000, covering sample areas across the whole moor, found birds in 97% of the squares surveyed, and the estimated total population was 13324 +/- 2032 pairs (Geary 2000). Because of different survey methods, looking at individual types of occupied habitat, an overall density figure per square kilometer, was not available for comparison. Even allowing for differences in the survey fieldwork techniques, it can be stated with conviction that Dartmoor's Skylark population has remained stable at a high level over the past 30 years, and may even have increased slightly. In view of the falling numbers elsewhere, this makes the population of regional, and almost certainly national, importance.

D'Urban and Mathew, in the late nineteenth century, knew that the breeding population left the moor in the winter, and this is still very much the case. By August very few birds are to be found on the moor, although some years there is an increase again in September as migrants pass through. Birds are then very scarce on the open moorland through the remaining autumn and early winter. It is quite probable that birds do not travel too far, but spend the time in lowland stubbles, or even agricultural areas on the edge of the moor. Flocks of 250 at Bridford on 18 November 1995, and 200 there on 10 December 2002, suggest this. Another reason for thinking that the bulk of the breeding population may still be close in winter is the fact that given the first mild days in February birds can be heard passing overhead, flying into the moor, and within a day or two song will be heard from the early males establishing territories. The main arrival back onto the breeding areas is later, with most birds arriving in early March, but by late March the lovely song of the Skylark is again to be heard everywhere.

SHORE LARK *Eremophila alpestris*
Very rare passage migrant.
The only accepted record of this species is a bird watched associating with two Reed Buntings near Avon Dam on 4 March 1995.

SAND MARTIN *Riparia riparia*
Scarce migrant breeder, and passage migrant.
Sand Martins are not well suited to Dartmoor. Their favoured habitat of low land, with high sandy banks lining slow meandering rivers does not exist on the moor, and yet come what may they have always seemed determined to preserve a breeding presence. This willingness to accept less than perfect breeding sites has led to a few rather bizarre observations over the years. The main alternative breeding habitat for birds has always been the clay pit areas of Lee Moor, where the soft, gravelly sides of the pits have proved the nearest thing available to their natural habitat. As breeding sites are not numerous they can be looked at individually.

Lee Moor and adjacent china clay area sites: A colony of eight to 10 nests was first noted here in 1962. Birds no doubt bred here from then on but were largely unrecorded. The next reference was in 1977 when 45 holes were noted to be occupied out of a total of 126. Almost annual counts in the late 1970s to late 1980s revealed an average of 37.5 occupied nests per year, although 61 were occupied in 1979. From about 1990 numbers began to drop, but rather than an actual reduction, this may have indicated that birds had moved to new colonies nearby at Crownhill Down, Wotter, or Portworthy Dam. This was certainly the case in 2001 when 21 holes were occupied, but a new colony had been located in the Torycombe Valley which had 30 active holes. In 2002 only seven pairs nested and the site was deserted by mid-August, two or three weeks earlier than the usual date. The Torycombe Valley site was not counted in 2002.

Cadover Bridge and the River Plym: Pairs were first known to be breeding in the river bank here in 1939. The colony no doubt changed its location on the river over the years due to bank erosion, but birds have stayed loyal to the general area ever since. Numbers have varied over the years from a maximum of about 14 pairs in 1976 down to four in 1994. This river section is a good feeding area, and the number of birds breeding can at times be made difficult to count by others being in the area, possibly feeding birds from the nearby Lee Moor colony. In 2000 a young bird was still being fed in the nest here on the late date of 1 September.

Powder Mills and the Cherry Brook: In 1939 a small colony was discovered on the Cherry Brook, nesting in the river bank close to Bellever Tor. Nothing was then recorded of them until 1965 when six pairs were found breeding on the Cherry Brook above Powder Mills. It had increased to 12 pairs in 1966, and eight pairs were present the following year. However, by 1970 only four pairs were counted, and this was the last record of breeding at this site, although in 1977 a small colony was located in a shallow sand pit by the main road, close to Powder Mills.

Prince Hall and the West Dart: In June 1961 six birds were seen over the West Dart near Prince Hall, and in 1963 a small colony of at least four pairs was found there, with about 25 holes showing that birds had been present in earlier years. Nothing then is heard of this site, although it no doubt still existed, until 1992 when 14 occupied nests were found near Prince Hall in a small roadside sand pit, and 16 birds were ringed. The colony was greatly reduced to six nests in 1994 and 1995, and in the next few years the sand pit was abandoned due to alterations making it no longer acceptable. Birds were still being seen in the area however, and in 2002 a small number of pairs were found nesting in the river bank, perhaps in the same location as 40 years earlier.

Bellever, Postbridge and the East Dart: Birds were recorded breeding in Postbridge on the East Dart during the 1920s, and in 1965 two pairs bred not too far away on the Walla Brook at Cator Common (Dare and Hamilton 1968). Nothing further was recorded in the area until 1996 when four occupied nest holes were found in a small roadside quarry on Riddon Ridge near Dury Farm. At about 340 meters this was thought to be the highest colony in Devon. Only a single bird was seen there in 1997, but in 1998 four or five of the 11 or so holes were tenanted, and young were heard in the nests. None were thought to be breeding 1999 to 2001, but in 2002 at least 10 holes were active with at least 18 birds in the area.

Other sites have held a pair or two for a short time, and amongst these, roadside sand pits and quarries have been used on several occasions. Instances of this have been the pair that nested near Dartmeet in 1952, and the birds attempting to use the Two Bridges car park quarry site in 1998. Above Okehampton one or two pairs nested on the Manor Brook in 1968, and a small colony was located there in 1972. The strangest location, however, must be a site used in Buckfastleigh. In August 1952 a pair were found nesting in a pipe in a wall overlooking the main road, and only three feet from the ground. In 1955 three pairs were nesting in similar drainage pipes along the same wall, and although nesting numbers were never again given, this strange site was used by birds until at least 1968.

Sand Martins are early to arrive in spring, with the first birds usually being seen in mid-March. The earliest bird on Dartmoor was without doubt the individual seen at Wrangaton on 27 February 1939. The main passage is usually in late March and early April, and is not usually well represented at Dartmoor locations. Burrator, however, has had good-sized migrant flocks on occasions, with at least 100 between 1 and 5 April 1994, and 150 on 18 April 1996. Birds leave the colonies in late August or early September, but autumn migration is very rarely noted over the moor. Twenty five were still at Portworthy Dam on 19 September 1995, but the latest were the five over Bench Tor on 1 October 2000.

SWALLOW *Hirundo rustica*
Migrant breeder and passage migrant.

The earliest ever Dartmoor record of a Swallow was the bird in the Postbridge area on 4 March 1943. Most years the first birds appear in the last few days of March or early April, with the main passage arriving from mid-April. There is still considerable passage in May, with the last of the local breeders arriving, and migrants going further north passing through. Dartmoor is no barrier for migrating Swallows and birds cross the moor on a broad front. Spring migration across the moor has been noted since at least 1938 when a north-northwest passage was watched over Ringmoor Down and Burrator on the 8 and 9 April. The numbers of over-moor migrants at this time of year are probably quite low, but surveyors out in the field for the Dartmoor Moorland Breeding Bird Survey in 2000 noted birds on several days between 19 April and 16 June. On 28 April about 100, in small flocks or singly, flew over Cut Hill between 0900 and 1600, all going northwest.

The breeding population on Dartmoor has never been fully assessed. It must be assumed that most farms have at least one breeding pair, but other than records from a few interested people, confirmation is not available. In the Postbridge area between 1956 to 1967 it was estimated that 50-75 pairs bred mainly at about 25 farms (Dare and Hamilton 1968). By the early 1990s numbers had decreased and the population then may not have exceeded 25-50 pairs (Dare 1996). The Harford Common Bird Census site has conversely shown an increase in nesting pairs since the early 1990s. Nine pairs in 1990 dropped to between five and seven in the years 1991 to 1995, but then increased to 11 in 1996, 12 in 1997 and 15 in 1998 and 1999. Nine pairs raised over 30 young at Moorlands Farm on the West Dart in 1992. A reduction in numbers at Ashburton was noted in 1992 and at Princetown in 1993, but information available is too little to make any useful comments on population trends. Pairs will at times nest in rather unusual places. A very exposed and isolated forestry hut was used in the 1960s at Soussons Down, and from at least 1995 to 2002 a pair have nested underground in an old mine shaft at Whiteworks. Two broods are not unusual, and three have been noted on occasions. At Welltown, Walkhampton a third brood of four fledged on 26 September 1997, and were last seen in the area in early October. At Lee Moor in 2002 a very late brood of three only left the nest on 3 October.

As can be seen from the above, local birds can still be present into October, but from late August many of the birds over the moor are migrants moving south. On 22 August 1989 about 200 birds were present at Two Bridges, and on the 27 August 1990 200 were watched going south over Soussons. A close flock of about 100 birds was watched going south over Hookney Down on 20 August 1999. They were quite disinct from the local family groups still feeding in the area. The main migration time is during September, and some larger movements have been observed over Dartmoor, usually towards the end of the month.

In 1982 a steady movement south was noted at Wrangaton on 20 September. At Okehampton between 7.30 am and 9.30 am on 24 September 1994 many thousands passed low heading into the southeast wind. Another huge movement was watched over Haytor Down on 14 September 2000, when at least 2000 moved south in the morning in flocks of 10 to 30. The main departure date noted at Huccaby in 2000 was 30 September. On this very fine day at least 200 flew east down the West Dart valley in a few minutes, and a steady stream followed afterwards throughout the afternoon. At the same locality in 2001 20 September proved to be the departure day, with birds taking advantage of several days of fine weather (Hibbert 2000 and 2001). Interestingly, at Lee Moor on the following day about 1000 birds lingered in the afternoon between 2 pm and 3 pm before moving south. That some migrants need to come down to roost in unusual locations at times was proved in 1964, when many were recorded roosting in bracken near the moor edge at Wrangaton on 27 September.

Late migrants have been seen several times in mid- to late October, and there have been six November records, the latest being one at Ashburton on 30 November 1994.

HOUSE MARTIN *Delichon urbica*
Migrant breeder and passage migrant.

The earliest House Martin was one seen at South Brent on 24 March 1996. This was extremely early, with the first arrivals most years being in the second week of April, and the main arrival being in late April and at times well into May.

As with the Swallow, the history of this species has not been well-documented on Dartmoor. What records there are of breeding numbers tend to show a population in decline. The most obvious case involves the bird's status in the Postbridge area. It was formerly a common summer visitor here with 150 to 200 pairs in about 15 colonies during the years 1956 to 1967. Challacombe Farm alone had 30 pairs in 1965. During 1991 to 1994 only 20 to 30 pairs could be found at eight sites. In 1991 only four farms were found occupied (Dare 1996). This was the best-documented decline, but on a smaller scale the Harford Common Bird Census site has shown a decline even more dramatic. Fourteen nests were found on buildings in 1987. By the next year it was down to eight, and had halved to four in 1989. Only a single nest was present in 1990, and none in 1991. In 1992 a pair returned, but there were no records thereafter. A census of the village of Horrabridge in 1998 found 48 nests, which rose to 65 the next year, and 60 in 2000. In 2001, without any reason being apparent, the number had dropped to 29. Numbers were also noted to be much reduced in Ashburton in 1992. Two regularly monitored sites, however, did show a more stable population over a period of years. The old Moorhaven Hospital buildings at Bittaford were counted annually from 1987 to 1998. This proved to be the largest single colony on the moor. Most years the total of nests was between 90 and 100, with a maximum of 122 in 1996, and a minimum of 53 in 1994. At the other end of the scale, the pairs in Bellever village were counted from 1994 to the present. Numbers have ranged between five and 10 over most of the years, and when they have fallen below this, as in 2000 and 2001, it has been because of recent repainting or alteration to the buildings. Other casual counts have shown 18 nests in Princetown in 1994, seven or eight in Walkhampton in 1998, with 47 at Halshanger Manor the same year, and 16 at Chagford in 1999.

There are fewer migrants seen over Dartmoor than with Swallow, especially in the spring, and even in the autumn there are never the huge movements noted at times for Swallows. Birds begin congregating in August, but it is usually mid-to late September before the largest numbers occur. On 18 September 1982, 300 were watched moving south over Haytor, and similar numbers have been seen over Okehampton on 3 September 1982, 18 September 1996, and 15 September 1997. The china clay pit areas of Lee Moor and Portworthy Dam see migrating numbers most autumns. In 1997, 200 were present on 11 September, and 300 on the 17 September. A heavy passage was noted on 21 September 2002 over Lee Moor with all birds heading southeast. Migration tails off quickly in early October, but there are usually a few birds still being seen up until mid-month. The latest seen have been individuals at Buckfastleigh on 1 November 1968, and at Okehampton on 22 November 1989.

RICHARD'S PIPIT *Anthus novaeseelandiae*
Very rare vagrant.
There are only two records of three birds. In 1994 two were found on Staple Tors on 14 October. They were associating with Meadow Pipits and were not found again the next day. The other record was of a bird flying south over Ugborough Moor on 22 September 2001. This individual was also observed with Meadow Pipit passage. It is probable that others pass south over the moor in autumn without being located.

TAWNY PIPIT *Anthus campestris*
Very rare vagrant.
The only record was a bird associating with *alba* wagtails in a field below Sheepstor on 26 August 1998. This bird was not relocated the next day.

TREE PIPIT *Anthus trivialis*
Migrant breeder and passage migrant.
In the late nineteenth century D'Urban and Mathew noted that in Devon the Tree Pipit arrived about the middle of April. This is still so, and on Dartmoor the first migrants arrive back most years between the 14 and 21 April. There have been occasional earlier April arrivals and even seven in March, the earliest being a bird at Burrator on 21 March 1988.

They require open habitat with scrubby ground cover, often bracken, and scattered trees for song posts. They have also been found in areas where gorse has taken the place of the scrub. Birds are widely distributed in suitable habitat around the moor-edge, and in the interior. The recording of birds annually from certain localities can be misleading, as totals appear to rise and fall regularly. Birds are located mainly by song, and if good conditions do not occur at the time of the counts, many birds will be silent and missed. Also many of the irregularities may result from insufficient observer coverage in certain years. A case in point is Burrator, where breeding pairs are regular in small numbers, but yearly figures have shown anything from five to 16 pairs/singing males in recent years. Young plantations are very attractive for Tree Pipits in their early years, but when trees get larger after about four years, the open ground between the trees is lost and birds vacate the area. There is then a period when the species can become quite scarce, only to come into its own again when the plantation is clear-felled and replanted many years later. Such a scenario occurred in the Postbridge area, where in the 1950s and 1960s 15 to 25 pairs were present mainly in young plantation and shelter belts (Dare and Hamilton 1968), and in recent years, after thinning and clearance of much of the earlier plantings, the number of pairs is now equal to or greater than the totals of 50 years ago. In 2001 at least 11 singing males were located in Bellever Plantation alone. Other areas with good numbers in plantation habitat have included Fernworthy, where up to 13 singing males have been found since at least 1979, in a large forest made up of a patchwork of tree blocks of varying ages.

In more traditional moor edge habitat, birds appear to be particularly well represented in areas around the east, south and southwest fringes, and less so in the north and northwest. The valley of the Avon from Shipley Bridge to Avon Dam had at least 10 singing males in 1978 and in 1987. The area around the edge of Piles Copse in the Erme valley had six in 1984, and eight in 1990, although numbers appeared to be down to three in 1997. There appears to be some loss of numbers in certain well-watched areas, although this may be a local thing due to habitat change. Two territories were found in the Harford Common Bird Census plot in 1990 and 1991, but although birds were present there was no breeding in 1992. In subsequent years birds were only encountered in 1996. Throwleigh Common produced up to six singing males between 1996 and 1999, but the number had fallen to two by 2000. One particularly interesting survey record was the 23 singing males found in the Dart Valley between Dartmeet and Holne Bridge in May 2002.

The 1979 RSPB/DNPA survey recorded 84 pairs, but the total was considered a considerable under-estimate (Mudge *et al* 1981). The Dartmoor Moorland Breeding Bird Survey of 2000 estimated the population to be about 249 +/- 48 pairs for the whole moor, and thought the population regionally important. This was a very significant increase on the 1979 figure. Comparison of the one kilometer squares covered by both surveys found Tree Pipits were present in 11 squares where they were unrecorded in 1979. Four of these squares contained forestry plantations, where clear-felling may have created habitat not available to birds in the earlier survey. Also the spread of bracken in many moorland fringe areas, may have provided additional suitable habitat (Geary 2000).

Spring passage migrants are very difficult to distinguish from breeding birds, and are thus rarely reported. The position in autumn is still difficult because of the lack of knowledge as to when breeding birds depart. Autumn passage does occur, however, and recent sightings away from breeding habitat, have included a bird that flew over Buckfastleigh on 21 August 2000, two that flew over there on 11 September 2001, and birds at Wootton, Holne on 3, 9 and 11 September 2002. On 3 and 11 September the birds were noted to fly south.

MEADOW PIPIT *A. pratensis*
Resident breeder and passage migrant. Altitudinal migrant with the breeding population moving off the high moor in the autumn.
The Meadow Pipit is the most numerous breeding bird on Dartmoor. It breeds in most grassland habitats, particularly blanket and valley bog, avoiding only heavily grazed areas with short sward where they are replaced by Skylarks. As with other common species, there is little recorded information from earlier days.

The abundance of the species has been well noted however in several surveys during the last 25 years.

In 1979 the RSPB/DNPA survey recorded birds in good numbers in virtually all habitats. They were recorded in 90.9% of all squares surveyed, and most of the squares in which they were not recorded only contained a small proportion of moorland. A rough estimate of the total population was 20,000 pairs. This figure gave an overall density of 43.7 pairs per square kilometer, which was considerably higher than certain other upland areas in Britain (Mudge *et al* 1981). The RSPB survey of 1992, looking at portions of the northern moor, showed a rather lower density of 26 birds per square kilometer (Chown 1992), but when a sample of the same area was looked at again for the High Dartmoor Sample Survey in 1997 the number of birds found had increased by 44%, suggesting that the 1992 survey may have shown a figure that was unrepresentatively low. The Dartmoor Moorland Breeding Bird Survey of 2000 estimated a total population for the moor of about 19,795 +/- 3028 pairs, which was very similar to the 1979 figure, and approached or just exceeded the 20,000 figure making Dartmoor of international importance for this species (Geary 2000). Three areas on the eastern side of the moor at Throwleigh Common, Manga Hill and East Okement Farm have been monitored for the BTO Breeding Bird Survey since 1993. Although there has been yearly variations, the results have been broadly in line with the densities noted for the main surveys. The rapid fall off in breeding numbers is noted as soon as the open moor gives way to agricultural land. Only one territory was occupied on the Harford Common Bird Census site in 1991, on agricultural land adjacent to open moor, with birds present but not on territory in 1992, and none since. In the off-moor parish of Bridford, a maximum of only three pairs were found in 1994, 1999 and 2000.

Birds tend to be later arriving in the spring than Skylarks, with the first week in March seeing the first song flights. The main arrival back onto territories takes place in mid-to late March, but even in April flocks of non-territorial birds can be encountered. These could well represent migrants stopping over to feed before continuing north. Examples of migrant flocking include 150 at Prewley on 22 April 1972, 80 at Cadover Bridge on 11 April 1998, and 40 at Down Tor the next day. After the breeding season family parties can be seen into August, but towards the end of the month numbers have usually decreased.

In mid-September the first migrant flocks pass through the area, with many coming off-passage to feed. In 1999 at least 200 birds were feeding amongst the dead bracken between Gutter Tor and Ditsworthy Warren on 12 September, and in 2000 at least 150 were on Ryders Hill on 16 September, with 300 in the Cadover Bridge/Brisworthy are on the 23 September. Smaller flocks can be encountered almost anywhere where there is sufficient cover, particularly in *molinia* and heather-rich areas. When visible migration was noted birds were usually moving south or west. During the second and third weeks of September 1997 a steady movement of small groups of 20 to 30 birds at the most were reported at Thornworthy, with most going south. These were never in tight flocks, but parties following each other fairly regularly (R. Waller *pers. comm.*1998). About 100 were watched flying west over Ringmoor Down on 19 September 1992. Occasionally birds have been noted moving in directions other than those expected. Between 1937 and 1945 late September movements of birds were noted at Wrangaton. These all moved in a northwesterly direction. As all were in the evening, they could have referred to birds going to roost somewhere close by, but as there were no returning birds in the mornings, they presumably moved on elsewhere. Flocks are still around the moor in early October, but many are now in moor edge localities, as the 150 at Wrangaton on 10 October 1986, the 160 close by at Moorhaven on 13 October 1987, and the 100 in one flock at Buckfast on 12 October 1952. Autumn and winter roosts on the edge of the moor are seldom reported, but one well-watched roost concerns birds near Yarner Wood. Birds were first observed flying in here in 1986, and numbers were normally about 50 to 70, although 118 were counted in on 24 October 1994. Interestingly, this roost is used from October to February or March. Presumably birds fly in from the agricultural hinterland to roost in winter, as by November there are very few birds on the open moor.

The only other roost described recently was near Leedon Tor, where on the 15 November 2001, about 30 birds were watched flying in from farmland in the Walkham valley to roost amongst the bracken at dusk.

As mentioned, by November very few are to be found, and their breeding haunts are almost completely abandoned. Occasionally singles or small groups will fly up, calling, from areas of heather or other cover, but the high moor is largely devoid of them until the following March. Where our breeding birds go in the winter is not known with any certainty. They may winter in France or Spain with other migrants from further north, or some may stay closer to home, taking advantage of the milder coastal regions, or even staying around the edge of the moor, as the 15 or so did that wintered at Buckfastleigh in the 2000/2001 winter. One notable exception to this, was a flock of 60 on Riddon Ridge on 15 February 1981. A very unusual occurrence that can be explained by the weather conditions at the time. The birds were moving ahead of depressions that were coming through, with snow hitting the region on 21 February. No doubt they had moved on by the next day. Another notable winter record was of 96 counted at Hamel Down on 15 December 2003.

WATER PIPIT *A. spinoletta*
Very rare passage migrant.
There have only been two records of this species for Dartmoor. The first was a bird present at Prewley on 6 and 7 March 1972, and the second a bird at Venford Reservoir on 24 and 25 October 2003.

YELLOW WAGTAIL *Motacilla flava*
Rare passage migrant.
Birds of the race *M. f. flavissima* that breed in Britain are birds of the lowlands, with quiet meandering rivers and water meadows. They rarely breed in Devon, and never have on Dartmoor.

The relatively few that have occurred on the moor have been passage migrants in spring or slightly more regularly in autumn. The earliest date was a bird that circled, calling, over Postbridge on 4 April 1987, then flew off to the east. Other spring records have been another at Postbridge on 30 April 1983, and one that flew south over Halshanger on 29 May 1991. Most of the autumn records are of ones and twos, usually seen at localities around the edge of the moor. A large party of about 20 was present at Yelverton on 24 August 1952. This record was probably linked with an unusually heavy passage that occurred in the Plymouth area at the time. Ten were watched with Pied Wagtails at Chagford on 21 August 1975. That birds do over fly the higher moor was shown by four seen over Harford Moor on 2 September 1982. They were flying south down the Erme valley with Swallows. Another record of two that alighted briefly at Riddon Ridge before flying on on 10 September 2000, also indicate an over-moor passage. In 2004, an interesting and unprecedented influx occurred in late August and early September. In an autumn not noted for its coastal passage, Dartmoor received records of about 50 birds, including a flock of 23, mostly juveniles, at Lower Hurston, Chagford on 24 August. This was the largest single flock ever recorded in the Dartmoor area. Most autumn records have been within this late August/early September passage period, but two were particularly late. A bird was seen at Prewley on 14 October 1970, and the latest were two on wires at Vitifer on 16 October 1960.

GREY WAGTAIL *M. cinerea*
Resident breeder, and altitudinal migrant with most birds leaving the high moor in winter.

There is evidence that the Grey Wagtail population increased and moved into new breeding areas during the mid-nineteenth century, and by the end of the century D'Urban and Mathew were referring to it as being particularly numerous on Dartmoor. Its status has remained much the same ever since.

It is a bird of fast-moving rivers and streams, and as such Dartmoor provides an abundance of suitable habitat. Breeding pairs are to be found on all streams, and as their prey is a wide range of insects and larvae, some of non-aquatic origin, they are little affected by the acidification of moorland water courses. Most records in the past have been of a casual nature, and although changes in population due to severe weather have been noted, it is only more recently that riparian surveys have shown actual numbers on given stretches of river. Most survey work has shown a pair for about every kilometer of river or stream.

A 4.4 kilometer stretch of the River Plym from Shaugh Bridge to Legis Tor has been surveyed annually for the BTO Waterways Bird Survey since 1984. Although there have been ups and downs, perhaps due to weather, the average over the years has been 4.9 territories, giving .90 territories per kilometer. A similar survey of 6.5 kilometers of the River Meavy from Burrator downstream since 1992 has shown an average of only 3.5 territories, giving .56 territories per kilometer. In 1993 the DNPA carried out survey work on 65 kilometers of the River Dart, including both East and

Grey Wagtail

West branches. They found 67 pairs of Grey Wagtails giving .97 pairs per kilometer. This figure was corroborated in 2002 when a 13 kilometer stretch of the Dart from Dartmeet to North Park Wood produced at least 14 territories, giving a similar figure of .93 territories/pairs per kilometer. A 5 kilometer stretch of the River Teign between Fingle Bridge and Chagford Bridge held 5 territories in 1982, thus giving 1 territory per kilometer, and a 2.75 kilometer stretch of the upper River Avon between Shipley Bridge and Avon Dam held 3 pairs in 1992, giving .92 pairs per kilometer. Apart from riverside surveys, there has been little long term assessment of population numbers. The main exception has been the Postbridge area, where 5 to 10 pairs bred between 1956 and 1967, and 9 to 10 pairs bred between 1990 and 1994 (Dare 1996). Also at the Harford Common Bird Census site one or two pairs have been regular since at least 1990. Thus we have a picture of a stable breeding population that declines on occasion after severe winters, but is well capable of regaining its former strength within a year or two.

Breeding territories are deserted in the winter, but birds may not go very far, perhaps only dropping down to moor edge farms, and village and town sites. They move back onto territories in March, and most pairs are re-established on moorland streams by the end of the month.

PIED WAGTAIL *M. alba*
Resident breeder and passage migrant. Also altitudinal migrant with most birds leaving the high moor in winter.
Pied Wagtails are widespread around the moor edge, and perhaps a little less so in the interior. The Dartmoor Moorland Breeding Bird Survey of 2000 found them

scarce and mostly near enclosures (Geary 2000), and although they will breed near water courses, they are very much more likely to be found near man-made sites in the breeding season than Grey Wagtail. In the Postbridge area, the population of 20-30 pairs in the 1956-67 period, was found mostly around farms and other habitation. It was hard hit by the severe winter of 1963 and lost perhaps over 50% of numbers. It did not regain its former numbers until 1967 (Dare and Hamilton 1968). In the early 1990s there were about 20 pairs in the area (Dare 1996).

In the BTO Waterways Bird Survey a section of the River Plym from Shaugh Bridge to Legis Tor has been covered since 1984. The normal number of pairs was usually 2 or 3, up until 1991 when 5 were found. Since then numbers have increased, with a peak of 9 in 1995. The average from 1991 has been 5.7 pairs, giving 1.3 pairs/territories per kilometer, very similar to the figure for Grey Wagtail. The same survey, carried out on the River Meavy for 6.5 kilometers downstream from Burrator from 1992, produced an average of 3.37 pairs, giving .52 pairs per kilometer, again very similar to the results there for Grey Wagtail. Three pairs were found on the River Avon from Shipley Bridge to Avon Dam in 1992 and 1994, giving identical figures to Grey Wagtail of .92 pairs per kilometer. Up to six territories have been found on the Harford Common Bird Census site since 1990, but the average has been between three and four. As with Grey Wagtail, birds mostly leave breeding sites in the winter for lower levels and easier feeding.

One interesting aspect of this bird's behaviour is its habit of roosting communally from autumn through to early spring. Many urban roosts can hold thousands of birds but on Dartmoor roost sizes are more modest. One of the first areas to be noted as holding a roost was Fernworthy Forest, where about 100 were noted in the plantation on 28 October 1962. A further 169 were watched flying into roosting trees on 31 December 1986. The maximum counts at Fernworthy were 200 on 4 November 1990, and 11 November 1996. Hennock Reservoirs have had a roost since 1980 at least, with the maximum count there being 100 on 15 February 1993. A roost at Trendlebere Down has been noted intermittently since 1983, and the maximum there was 257 birds counted flying in from the east on 24 October 1994. At Buckfastleigh on 13 March 1994, 184 were roosting at Kilbury Sewage Works. Of the other roosts perhaps the most inportant recently has been the clay pit site of Lee Moor. This site is particularly interesting because the roost builds up earlier than elsewhere with numbers up to 100 in August and September, with birds still coming to roost in spring, sometimes as late as the first two weeks of May. The maximum there has been about 350 on 21 October 2001.

Some of the autumn and spring birds at roosts could be migrants of the nominate race *M. a. alba* known as the White Wagtail. As the races are rather difficult to differentiate without a good view, it is possible that White Wagtails may pass across Dartmoor unnoticed. There have only been eight definite records of White Wagtail on the moor since 1961. Five have been in spring, early April to mid-May, and three in autumn, mid-September to early October. The only locality to have more than one record is Meldon Reservoir, with two on 4 October 1991, and singles on 20 April 1994 and 19 April 2002. One on Ugborough Moor on 5 April 1996, was unusual in that it was the only bird to be seen in open grass moor habitat.

WAXWING *Bombycilla garrulus*
Very rare winter visitor.

This winter visitor from Scandinavia arrives on the east coast of Britain in varying numbers most years. Most stay on the east coast, but in years of exceptional irruptions some birds will make their way westward in search of food. A few will then usually reach Devon, and of these, on occasions, one or two will wander to Dartmoor. Being heavily reliant on crab apple, cotoneaster berries and the like for food, they appear in gardens offering an acceptable food source, and when this is expended they quickly move on.

The first Dartmoor record would appear to be the two seen on cables near Haytor on 13 January 1958. The next records were in 1965, which was a major irruption year. Nine were present at Throwleigh on 6 January and up to 20 were seen at Buckfast and Tavistock from 25 November. The next was a bird at Okehampton Station on 21 December 1983.

One was watched on an elder in an Ivybridge garden on 24 December 1988, and 1996 saw one at Yelverton on 9 February. Although possibly outside the Dartmoor area, five were seen at Tavistock on 9 March the same year. The most recent sighting was of one in a Yelverton garden feeding on crab apples, on 2 February 2003.

DIPPER *Cinclus cinclus*
Resident breeder.

The Dipper, along with the Grey Wagtail, is the characteristic bird of swift-flowing Dartmoor rivers and streams. Birds inhabit all major river systems on the moor, and most of the smaller streams. They will breed on any stretch of fast-flowing water, as long as their food supply of mainly aquatic invertebrates and their larvae is sufficient, and as long as there are enough nesting sites. Natural nesting sites are used, but the majority of pairs breed on man-made structures, especially under bridges.

Tony John has studied Dippers on the catchment areas of the Plym and Tavy since the late 1970s. Over a period of rather more than 25 years, the population in the study area has been found to be basically stable. Some of the study area has included stretches of both rivers off the Dartmoor area, but details are given for interest's sake. The number of broods raised has varied between a high of 17 in 1994, when three pairs, rather unusually for Dartmoor, had second broods, to a low of seven broods in 1981 when many nests were destroyed by unseasonal late April snow. The average number of broods annually has been 12.

The number of young per brood has averaged 3.5. In addition to this study, stretches of the Plym and Meavy have been surveyed for the BTO Waterways Bird Survey. On the Plym, a survey of the 4.4 kilometers between Shaugh Bridge and Legis Tor has produced an average of 2.13 pairs, giving .48 pairs per kilometer. On a 6.5 kilometer stretch of the Meavy downstream from Burrator an average of 4.85 pairs has been found, giving 1.34 pairs per kilometer. In 1993 the survey of 65 kilometers of the River Dart, including East Dart and West Dart, for the DNPA, found 43 pairs, giving an average of 1.51 pairs per kilometer. Reservoir sites can also prove attractive, with a maximum of five pairs in the Burrator area in 1992, and four pairs near Meldon in 1990. Although bridges, etc., are most popular as nesting sites, occasionally a more unusual location will be used. About 1993 a pair nested on the East Dart near Bellever in a tree nest about eight feet up over the river in a large thorn bush. The young were successfully fledged (G. Weymouth *pers. comm.*1996). Many sites are traditional, though. A riverside site at Dartmeet had been used every year for at least nine years up until 2000, and no doubt many of the bridge sites can claim decades of use, if not more.

After fledging, the young disperse to other river systems, although they rarely go more than 10 kilometers from their natal area. Females tend to go furthest. Their choice of habitat, when young, is not perhaps always the best. On 14 and 15 July 1999, a probable juvenile was seen at the bottom of a working clay pit at Lee Moor. The scene was repeated in 2001 when a juvenile was again at the bottom of a working clay pit at the same locality on 8 and 10 July. What was probably the same bird was later found dead. Not all young are so unfortunate, however, and one bird ringed at Lamelgate on Bodmin Moor on 14 May 1970, was re-trapped at Wilminstone Quarry, Tavistock, just outside the Dartmoor boundary on 25 September 1978. At 8.3 years this was the oldest Dipper recorded, a record that probably stands today.

There is only one area where the population from the mid-twentieth century can be compared with 40 or so years later, and that is in the rivers and streams around Postbridge. During that period there was something of a decline. In the years 1956 to 1967 five to 10 pairs were found on all the main waters, with birds exploring streams to near their sources, even within plantation areas (Dare and Hamilton 1968). By the early 1990s this number was down to about five pairs (Dare 1996). It was reported that Dippers disappeared from the West Webburn with the arrival of Mink. Perhaps the increasing acidification of the streams in the area, due to the maturing conifer plantations, also played a part in this decrease.

In the nineteenth century there was the belief that Dippers moved off the moor to lower ground in severe winter weather. This may be partially true, but there is no proof that birds move far, and they do not appear to be affected by severe winters. The number of pairs breeding around Postbridge after the Arctic 1963 winter, had not declined. An instance of a bird moving between waterways at this time of year, was the record of a Dipper flying over the moor at Merripit Hill at dusk on 25 November 2001.

WREN *Troglodytes troglodytes*
Resident breeder.

Widespread and adundant from the 'in country' to the high tors, the Wren has been able to adapt to most Dartmoor habitats. What it unfortunately has not been able to adapt so well to are the severe Dartmoor winters. Most hard winters take their toll, and the worst of them have really disastrous effects on the population. In the late 1950s the population around Postbridge was estimated to be about 250 to 500 pairs. They were known to have been hit badly by the 1946/47 winter, and in the winter of 1962/63 they were almost completely wiped out, with mortality probably near to 99%. Only four singing males were located in the area during the 1963 breeding season. There was a slow increase in numbers in 1964-65, a marked increase in 1966, and a full recovery by 1967 (Dare and Hamilton 1968). The species was found to be abundant here in the early 1990s, with numbers probably increased since earlier years due to mild winters and the maturing of the plantations (Dare 1996).

The monitoring of the Harford Common Bird Census's 76.7 hectare farm site between 1988 and 2000, produced an average of 27 pairs/territories per year, with a maximum of 35 in 1989, a 20% increase on the year before, and a minimum of 17 in 1997. Dunsford Wood Nature Reserve, on the eastern edge of the area, has shown fluctuations caused by the severity of winters. After a cold winter in 1991 only 11 territories were located, but within two years a full recovery had been made and 28 territories were found. This was followed by a bumper year in 1994 when the total rose to 32, the highest since 1980. Further, but less sharp, declines and increases were noted in the late 1990s. The BTO Breeding Bird Survey coverage of a one kilometer square at Throwleigh Common gave a total of 20 in 1999, maximum 21 in 2000, and an increase to 31 in 2002. Recent survey work for the Devon Atlas Project in 2002 has given some interesting counts, with 16 singing males along a 2.5 kilometer stretch of the River Swincombe, 17 singing males along the West Dart river, and adjacent farmland north of Hexworthy, mainly in tetrad SX6573, and 47 singing in Bonehill tetrad SX7276.

In the survey years for the Devon Tetrad Atlas 1977 to 1985, it was found that Wrens bred on Dartmoor up to at least 400 metres above sea level (Sitters 1988). Records since have shown that there really is no limit, and if there is sufficient cover birds will breed on the highest tops. In 1996 two were found in Steeperton Gorge at 430 metres, and in 2002 one was singing in the same area at Knack Mine about 450 meters. The record that proved that altitude was in itself not a problem, however, was the nest that was found at Fordsland Edge, near High Willhays in the mid 1980s. This was at 580 metres above sea level, and only 41 metres short of the highest point on Dartmoor and in southern England.

DUNNOCK *Prunella modularis*
Resident breeder.

There is evidence that this unobtrusive species has declined on Dartmoor over the past 40 or so years. As with many other common species, little notice was

146

taken of them when they are plentiful, so comparisons in many cases cannot be made. However, during the years when the Postbridge area was being annually surveyed in the 1950s and 60s, an estimate of 100 to 200 pairs were thought to be present. A marked decrease was noted after the 1963 winter when mortality was probably more than 50% (Dare and Hamilton 1968). Its earlier status was no doubt regained in a few years, but a survey of the same area in the early 1990s found only 15 to 20 pairs (Dare 1996). This drastic decline was put down largely to loss of habitat, through the maturing of the conifer plantations, but although this is no doubt mainly correct, other areas were also noting a loss, where a change in the habitat was not an issue. In the spring and summer of 1993 the Dartmoor Study Group conducted a survey of the three areas of relict pedunculate oak woodland on Dartmoor, Wistman's Wood, Piles Copse and Black Tor Copse. One of the surprises of the survey was that no Dunnocks were found. Although not prime habitat, three had been noted in Piles Copse in April 1947, and several had been seen in the area of Wistmans Wood in 1970 (Niles 1971). The Common Bird Census records from 1972 to 1996 show declines nationally of 40% on farmland and 41% in woodland (Mead 2000). As yet these national declines, mirrored no doubt on Dartmoor, are unexplained.

At Harford Common Bird Census site the Dunnock population, although small, remained stable from 1987 to 2000. Although 1995 did not show any breeding pairs, in most other years three or four territories were found. What would appear to be a high total of nine singing males was counted in the Prince Hall area in May 2000. In 2002 field work for the Devon Atlas Project found two singing males on either side of Bench Tor, but none in the Dart woodlands below. Birds were noted to be scarce in these oak woods. A 2.5 kilometre stretch of the River Swincombe, between the intake and Wydemeet Bridge, had five territories, and at least six pairs were located on Yennadon Down. The Bonehill area, tetrad SX7276, had a total of 21 singing males.

Birds take readily to gorse scrub, and because of this can be found quite deep into the moor. In 1992 birds were found in May on Smallbrook Plains, where they had not been recorded during the fieldwork for the *Devon Tetrad Atlas*. A pair was feeding young in western gorse at Merrivale Warren, below Great Mis Tor, in May 2000 and 1 December 2002 a bird was watched in gorse in open moorland, on the East Dart by the Beehive Hut.

ROBIN *Erithacus rubecula*
Resident breeder, and possible passage migrant and winter visitor.
This bird is widely distributed and common in all suitable habitats on the moor, from plantation and broad leafed woodland to copses and farmland, but very few estimates of population numbers have been made. In the Postbridge area during the 1950s and 1960s the population was thought to be about 100 to 150 pairs. The severe winter of 1963 hit them hard and mortality of about 50% was noted (Dare and Hamilton 1968). The numbers were soon restored however, and the early 1990s found them

most numerous in the conifer plantations, where an increase on the 1960s numbers was thought probable (Dare 1996).

Elsewhere, peaks and troughs, perhaps just localised, have been noted. At Harford Common Bird Census site 19 territories were found in 1987, only to be increased by 50% to 28 in 1988, and a further 57% to 44 in 1989. By 1990 the 45 territories made them the commonest species present, and in 1991 and 1992 a peak of 50 was reached. The 1993 figure was down to 40 territories, and numbers then stayed about the same until 1997 when a low of 26 was encountered. Numbers then increased again, until the figure of 45 was again reached in 2000. On the eastern side of the National Park the Dunsford Woods Nature Reserve showed a similar pattern, with highs and lows, that in the main corresponded with Harford. At Dunsford 42 pairs in 1990 was the highest number since 1980, but numbers had dropped by 60% to just 17 in 1991, suggesting that the local population had not coped well with the severe weather of February that year. This low was not noted at Harford. Numbers picked up rapidly until a further drop to 16 pairs occurred in 1997, but again numbers were back to normal in a year or two, and another high of 41 was reached in 2000. Survey work for the Devon Atlas Project in 2002 has shown some good counts with 48 singing males being found in the Bonehill tetrad SX7276, and at least 14 pairs on Yennadon Down.

The presence of migrants in the autumn and winter is somewhat tentative, but there have been indications that this could be the case. In 1986 at Moorhaven Hospital, Bittaford, 30 were counted on 8 October, where the norm was 8/10. In 2001 exceptionally high numbers were noted on the eastern edge of the moor in the autumn and second winter period, and that same year a bird was found amongst rock clitter on the northern moor at Wild Tor Well on 15 December, all suggesting that immigration does occur.

NIGHTINGALE *Luscinia megarhynchos*
Very rare summer visitor, breeding occasionally.
The Nightingale has never been regular on Dartmoor. As Devon lies at the western extremity of the bird's British range, the county population has always been small and vulnerable. It was at its height during 1930/35 and this was mirrored by slightly more Dartmoor records than usual in the period.

D'Urban and Mathew noted a bird being heard near Horrabridge in April and May 1882, and two or possibly three pairs in coppice near Ashburton in 1888 and 1889, after first being heard at the site in 1887. A bird was recorded at Bridford in 1918, and the species was found there again in 1930. On 1 May 1924 W. Walmesley White heard two singing on the northern slopes of Dartmoor at about 700 ft. above sea level, perhaps Skaigh or Okehampton Park area. Birds were heard at Hennock in 1930 and 1932, and in 1936 two were singing in the Dart valley at Double Dart on 12 June, with one still present on 14 June. This is the only Dartmoor record away from moor edge areas.

There were no records in the 1940s, but a survey by the Devon Bird Watching and Preservation Society in 1950 found a singing male at Buckfastleigh, where a bird was also present in 1951. A bird was also heard at a site two miles south of Christow village in May 1951. About 1960 a singing male was located on the western edge of the area, on the Tavy/Walkham confluence at Double Waters. There followed 28 years without a Dartmoor record. By the late 1980s the breeding range of the Nightingale had contracted greatly in Devon, and there were signs that its main stronghold in the Bovey Basin, on Chudleigh Knighton Heath, was beginning to loose numbers. It was a little surprising then that a pair was proved to breed successfully in 1988 at Ilsington, and what was probably a migrant was seen at Okehampton on 29 August..

Breeding ceased at Chudleigh Knighton Heath in 1993, and it could well have been expected that Dartmoor records were a thing of the past. However, during the Dartmoor Moorland Breeding Bird Survey of 2000 a singing male was found during survey work on Rhôs pasture in Tor Valley near Throwleigh (Geary 2000).

BLACK REDSTART *Phoenicurus ochruros*
Scarce passage migrant, and winter visitor.

The distribution of the Black Redstart in Devon is mainly coastal for passage migrants, and wintering birds. The birds that wander to Dartmoor are a very small proportion of those passing through the county. It is a species that has an affinity with man-made structures, and even on Dartmoor this is apparent, with most records coming from towns or villages around the moor, or industrial buildings, reservoir dams and the like.

About 20 birds have been noted in the spring migration period on Dartmoor. The dates of the majority correspond with coastal passage, with most in the last week of March and first week of April. The earliest record is of a male at Lee Moor on 11 March 2002. This could have been a bird that had wintered locally, although males are known to lead the way in March passage. Unlike autumn and winter records, birds in spring are seen in mainly natural habitat. Vitifer has had records on 3 April 1969, 28 March 1982 and 31 March 1997, Grenofen on 20 March 1983 and 1 April 2000, with Tavy Cleave, Haytor, Lynch Tor, Meldon Valley, and Sampford Spiney all having singles over the years. Birds have even got to the highest tors with a male on Yes Tor on 21 April 1987, and a pair at Fur Tor on 4 April 1997. There has never been any suspected breeding, but there have been two records of birds seen well into the breeding season. One was watched on the Avon Dam on 15 May 1983, and a male was seen on the spoil heaps by Haytor Quarry on 18 May 1997.

The first date for an autumn migrant is 30 August when a bird was recorded at Lee Moor in 1997. This was probably a dispersing juvenile from a natal site not too far away, but the main passage, of what are probably continental birds, is during late October and especially early November. This is the period when most Dartmoor records have occurred. About 90 have been recorded, but it is certain that many more must pass through unnoticed. Because of the good feeding, necessary to gain weight and strength before proceeding further south, birds will congregate in urban areas at this time. Buckfastleigh, Okehampton and Chagford have had up to four birds stopping over for a few days, and South Brent and Tavistock have had singles. Other man-made sites have included the water treatment works at Prewley, where up to three birds were regular in the autumn from 1969 to at least 1977, and the hospital complex at Moorhaven, Bittaford, where up to three were recorded from 1985 to1990. Birds in more natural surroundings at this time have included singles at Roos Tor in 1971 and 1977, Fox Tor Mires and Powder Mills in 1980, Wild Tor Well in 1996 and Hookney Down in 2000.

By mid-November most of the migrants have passed on, and the birds that are then present are birds that stay to winter. The main wintering site for birds is without doubt Buckfastleigh. In recent years three or more birds have been in the area most winters. Birds are present from mid-to late November and stay until February or early March. Okehampton, Ivybridge and Chagford have also had wintering individuals, and occasionally other more exposed moorland areas have had records, the most extreme being the bird at Cut Hill on 12 January 1981.

REDSTART *P. phoenicurus*
Migrant breeder, and passage migrant.
The story of the Redstart as a breeding bird on Dartmoor really began in the early 1950s. Before then records were very few. D'Urban and Mathew knew them as being common in east Devon and possibly the north, but remarked that they were scarce in the south and west of the county, although they were supposed to be spreading westwards. They made no specific references to Dartmoor.

Birds were noted in Lustleigh Cleave and Drewsteignton in June 1934. Individuals were seen at Lustleigh again in 1936, and at Hennock in 1941. There were two breeding pairs at Holne in 1942, and by 1949 an increase was noted with birds at Fernworthy, Moretonhampstead, North Bovey and Manaton. During the early 1950s birds were found in many woodland and moor edge sites on the east side of the moor. There was every indication then of a rapid expansion. In Yarner Wood five pairs were breeding in 1954, where none had been found in 1951 (Page 1992). The population there had increased to 12 pairs by 1956. Colonisation of suitable habitat in the central and western parts of the moor followed rapidly. The West Webburn and Walla Brook valleys were found to be holding surprisingly good numbers in 1956, with nearly all copses having a pair or singing male. Birds were located at 16 sites, and bred at 10. In 1955 a pair bred at Burrator for the first time. On northern parts of the moor birds were also appearing, with singing males at Doe Tor Farm and Black Tor Copse in 1956.

In 1957 the Devon Bird Watching and Preservation Society launched a Redstart Enquiry to collect more information on this expanding population. The Enquiry ran until 1960, and the results confirmed that Redstarts were doing very well. The numbers at Burrator increased to five pairs in 1957, and nine pairs in 1959. The area around Bovey Tracey, including Yarner Wood, held between 10 and 15 pairs, and the Postbridge area had 18 to 22 pairs by 1958. Birds were turning up each year in new areas, with four pairs being found around Hexworthy in 1957, and others at Cornwood, Widecombe and Christow. By the end of the Enquiry it could be stated that breeding was occurring over almost the whole of Dartmoor, in suitable upland copses and around farms (Ellicott and Madge 1961).

The consolidation of the population in the area, with most suitable habitat occupied, carried on throughout the latter decades of the last century. Thirty to 35 pairs were in the Postbridge area by 1967 (Dare and Hamilton 1968), but here the population stabilised, as the same number of males were recorded in the spring of 1990 (Dare 1996). At Burrator 21 pairs bred in 1981, and 25 the next year. A count of singing males here in 1996 found 23, so presumably here too the population had stabilised. The fact that even during the 1990s, there was still some room for expansion was proved at the Harford Common Bird Census site, where between three and six territories had been the norm from 1990 to 1996. Astonishingly in 1997, the total increased to 10, with 13 the next year, and no fewer than 15 in 2000. The fortunes of this species have not been totally an upward spiral since the 1950s however. During the mid-1970s many sites saw a marked reduction in breeding numbers for a few years. The reason for this was probably the drought in the Sahel zone of the Sahara, the wintering area for the species, although Redstarts are known to be naturally prone to fluctuations in population. By 1979, however, numbers were well on track for a full recovery.

The casual records published each year in the reports of the Devon Bird Watching & Preservation Society and the Dartmoor Study Group do little to establish an estimated Dartmoor population for this widespread species. Numbers given are always under estimates, and even when certain areas are worked well it is usually difficult to be sure of the actual ground covered. The 1979 RSPB survey of the moor found only 31 pairs, but the survey methods were clearly not suitable to census this species, as the total was unrealistically low. The DNPA Dartmoor Moorland Breeding Bird Survey of 2000 found them on, or close to, the moorland fringe, closely associated with the tree line, including plantations. They were widely distributed along the southwest and westerly edges of the moor. The total population was estimated to be about 268 +/- 166 pairs, but as upland oak woodland was not surveyed, the actual total could well be higher (Geary 2000).

The first birds, usually males, arrive back onto breeding territories in mid-April most years. Particularly early dates have been birds at Broad Down on 31 March 1974, Widecombe on 1 April 1991 and Yarner Wood on 4 April 1994. The main arrival takes place from late April, through early May. On Dartmoor, birds are heavily reliant on stone walls for nesting sites. Of the 25 pairs breeding at Burrator in 1982, only one nested in a tree hole, the rest were in walls. Birds will accept suitable nest boxes, but it appears that they have only been successful at Yarner Wood, and even there the success rate is variable. One feature of the bird's acceptance of less than prime habitat, has been the frequency in recent years that territories in conifer plantations have been found. This was almost unknown in the 1970s.

Most adults and juveniles leave the breeding grounds during August, and birds that occur later are probably migrants from further north, moving through. Passage birds have occurred several times in September but unusually late were the records from Moorhaven on 5 October 1987, and Ingra Tor on 20 October 1973.

WHINCHAT *Saxicola rubetra*
Migrant breeder, and passage migrant.
In the 1933 *Annual Report* of the Devon Bird Watching and Preservation Society, a comment was made by Douglas St. Leger Gordon that Whinchats were tolerably plentiful, but local, in distribution. The same comment could well still apply today. Many early statements on the bird's status mention it as a localised breeder.
It has always been associated with areas of light grazing and rank vegetation, especially around valley mires. Bracken can be important in territories, as can extensive heather in others, notably the general area of Warren House Inn. However, many apparently suitable areas are untenented, or at best only hold single pairs.

This localised distribution has produced several traditional Whinchat areas. The Rev. R.A. Julian found one or two pairs annually at Cadover Bridge and published a note to the effect in *The Naturalist* of 1851. Down through the years to the present time one or two pairs can usually be found in the same area. Another example of an area that has always been attractive is the valley of the River Avon between Shipley Bridge and the surrounds of Avon Dam. Pairs have been noted here since at least 1960 in varying numbers. In the last decade three or four pairs have been annual, but before this numbers were higher, with six pairs in 1984 and 1987, nine pairs in 1988, and no fewer than 14 singing males found on 10 May 1978. Certain areas attract a dense population, only to lose most again when the habitat changes. In the 1950s and 1960s many pairs were found breeding amongst young trees and vegetation of the new plantations at Bellever and Soussons. This had probably been also the case some years earlier at Fernworthy, but is not so well documented. In 1957 12 pairs were found breeding in about 40 acres of Soussons Plantation, with four pairs of Stonechats. The estimated population for the whole area around Postbridge during 1956 to 1962, including the plantations, was 50 to 75 pairs (Dare and Hamilton 1968). As the conifers matured the plantations rapidly became unsuitable, and the population reduced to about 25 to 35 pairs in the early 1990s (Dare 1996). Close by in the heather covered slopes of the Warren House Inn area, including Vitifer, Headland Warren and Birch Tor, though noted as a breeding area from at least 1947, it was not until the 1970s that numbers increased greatly. From 1976 ten pairs were normal with an increase to 12 to 15 in 1994 and 1995.

Helpful though casual records are in estimating the numbers in specified areas, they are only of limited use in trying to assess the population of the whole moor. Since 1979 several professional surveys sponsored by the DNPA and RSPB have taken place. The Whinchat population declined in southern Britain in the 1970s and 1980s, all but disappearing from many lowland areas (Gibbons *et al* 1993). Dartmoor was spared this and the population was found to be stable, or even in places increasing. The 1979 the Breeding Bird Population Survey found territories well distributed on the southern and eastern sides of the moor, mainly in the lower and middle altitudes with a peak at 1100-1200 feet. There were few records from northwestern areas, and they were found to be largely absent from the high plateau. A total of 208 pairs was recorded during the survey (Mudge *et al* 1981). A sample survey in 1992 (Chown 1992) found a substantial increase in areas of the northern moor surveyed, although it was thought that the particular areas may have been underrecorded in 1979. The Dartmoor Moorland Breeding Bird Survey of 2000 confirmed the substantial numbers on the moor, and suggested that the population may have been in the region of 577 +/- 222 pairs, thus giving up to a 276% increase over the 1979 total (Geary 2001). This figure was perhaps too optimistic, but the population is very important and almost certainly must cross the threshold of 140 pairs at one site, making it of national and probably international importance.

Whinchats arrive on breeding territories from mid-April, with most arriving in the last week of the month, and early May. Occasionally birds that are perhaps passage migrants, as against breeding stock, are seen earlier. Examples are the individuals at Avon Dam on 31 March 1975, and Yarner Wood on 1 April 1999. Autumn passage begins at the end of July, and in August family parties can be seen together prior to departure. Such a party was the 20, including about eight juveniles, at Holwell Tor on 23 August 1983. The somewhat protracted migration period stretches through most of September, with the possibility then that some may be continental birds. Very occasionally a daytime movement of birds may be seen, as with the small numbers going southwest near Harford for about two hours around midday on 9 September 1960. Only a trickle of birds appear in late September, but even in the first week of October a few are sometimes still around. Exceptionally late migrants were the two at Fernworthy on 28 October 1996, and the single at Hare Tor on 7 November 1998.

STONECHAT *S. torquata*
Resident breeder, and possible passage migrant. Also a partial altitudinal migrant, with many, but not all birds, coming off the high moor in winter.
Up until the latter decades of the twentieth century, the Stonechat was rather taken for granted on Dartmoor. Annual Reports of the Devon Bird Watching and Preservation Society make little mention of them, and there is no way of estimating the population of particular areas, let alone the moor as a whole.

The first survey work of real assistance was carried out in the Postbridge area between 1956 and 1967. The population here varied according to the severity of the previous winters, from one or two pairs, as after the severe 1963 winter, to 10 to 15 pairs. They were noted to be mainly summer visitors, present early March to September, with very few wintering, even in mild winters (Dare and Hamilton 1968). A survey of the same area during 1990 to 1994 also estimated a population of 10 to 15 pairs. Many birds leave the moor in cold winter weather, and move to lower ground, notably on the coast. This does not stop the effects of severe weather totally however, and mortality can be high. Unfortunately, the DNPA/RSPB survey of 1979 took place after a severe winter, and consequently the numbers located were rather low compared with later surveys. A total of 157 pairs was found, but there were instances where no birds were located in previously well-stocked habitat. An example of this were two one kilometer squares on Roborough Down which did not yield a single bird. This same area was surveyed by the writer in 1974 for the Devon Breeding Status Survey and 12 pairs were found (Sitters 1975). A count here in 1996 found 20 pairs, and 17 to 20 were confirmed again in 2000.

The Stonechat, always primarily a bird of gorsey heaths, did very well on Dartmoor in the 1980s and 1990s, and spread onto other habitat, notably heather moorland. The Dartmoor Moorland Breeding Bird Survey of 2000 noted this, and also a huge increase in numbers since 1979. The population for the whole moor was estimated to be about 1612 +/- 397 pairs (Geary 2000). This was a massive increase, and although the low 1979 figure was obtained in a poor year for the species, the 2000 figure does appear to be unrealistically high. No doubt future surveys will confirm or otherwise. However, there is no doubt that Dartmoor has great significance for

Stonechat

154

Britain's Stonechat population, and as it certainly exceeds 100 pairs at one site, it is of international importance.

Most males are back holding territory by early to mid-March. Some, especially in mild years, may have remained close to the breeding territory throughout the winter. Others may not have gone far, as with the 10, mostly males in bright breeding plumage, found feeding together in a kale field near Ashburton on 15 March 1975. The breeding season is protracted from late March into early August, with pairs regularly producing two or more broods. Extended family parties are seen in early autumn, and at times the size of these parties suggest more than one family feeding together. On Roborough Down on 20 August 2000 28 birds, mostly juveniles, were counted. Birds will take advantage of atypical habitat for feeding occasionally at this time of year, perhaps after a good breeding season and all the attendant pressures to find sufficient food. At Huccaby in early August 2001 juveniles were found to be collecting in an uncut grass valley bottom field to feed in the early evenings. They were absent during the day. They fed quietly without any calls. Birds were noted from 5 to 12 August, and the maximum seen was at least 14 on 9 August (Hibbert 2001). The same year about 20 birds, mostly juveniles, were noted at Ingra Tor on 17 August. Pairs and dispersing juveniles are seen throughout September, but by October birds are beginning to thin out. A high count of about 40 birds, comprising seven family parties, were seen in the Ingra Tor/Routrundle area on 26 October 2002, but by 13 November only two were present. By November most birds that are going to leave the high moor for the winter have left, relocating to coastal areas and possibly going further south. In 1990 a bird that had been ringed as a nestling by the Warren House Inn on 3 June, was controlled at Canete la Real, Cadiz, Spain on 15 October. The birds that have decided to stay the winter, normally adults, then take up a territory which is usually larger than their breeding territory, although it may well be in the same area. They remain until at least early March, when the males get territorial, and breeding sites are again occupied.

A footnote must be added about certain males that appear on the moor in spring. These birds are stunning, and look almost black and white at a distance. Their upper parts are very dark, almost black, and they have large white neck and wing patches. Their underparts are an apricot colour with a small triangular area of firey orange on the upper breast. They have white rumps, and heads that look almost black. They are totally different to the usual resident males. They appear mostly quite late in the spring in May or even June, and sing and display to resident females, who no doubt by then have had one brood. They have occurred over many areas on the coast, as well as on the moor. No explanation has been given as to their occurrence, but it would seem that they must be migrants from the continental race *S.t. rubicola,* which is brighter than the British and Irish race *S.t. hibernans.* They are known to intergrade with *hibernans* in northwestern France (Wernhan *et al* 2002), and perhaps birds have been crossing the Channel to interbreed with our local population for many years without much notice being taken.

WHEATEAR *Oenanthe oenanthe*
Migrant breeder, and passage migrant.
Dartmoor has always had the largest population of Wheatear in southern Britain. Recent surveys have shown that numbers have increased dramatically overall in the last 25 years, and yet individual sites have seen declines.

There is no doubt a natural fluctuation of population over a number of years, as with most species, but there are two constants that must be present in all breeding habitat for Wheatears. There must be ample stone walls, rock clitter, etc., for nesting sites, and there must be good areas of short grass sward for feeding. The large numbers of stock grazing, or over-grazing, the moor in recent years has greatly benefited the Wheatear population. This is particularly evident on the south and west sides of the moor, which have always been favoured areas. The sites that have seen some decline have probably had a habitat change, with local reduction of stock allowing the growth of longer sward. The most recent survey showed that total absence from a once-occupied area was found most often in heather moor, where the heather could have got too high to be suitable (Geary 2000). Casual records do not normally show enough constant recording to be very helpful, but a few areas have been looked at almost annually for a few years, and these show findings at odds with the general trend. A one kilometer square centred on Routrundle in the Walkham Valley, was surveyed yearly from 1993 to 1998. Seven pairs bred in the area in 1993, with seven pairs breeding again in 1994, although in that year there were up to nine more males/pairs in the area that possibly bred. The confirmed breeding number dropped dramatically to three in 1995 with the same number in 1996. Only two pairs were found in 1998. The southern edge of Wigford Down, from the Dewerstone to Cadover Bridge, along the walled edge of the Cadworthy enclosures, was surveyed from 1992 to 1995. In 1992 eight pairs were found along this stretch. This reduced to five the next year, and was down to a single pair in 1995. The fact that numbers have not increased greatly in this area since the latter date, would appear to indicate a habitat change, rather than any natural fluctuation. These are just two instances of localised declines in the mid-1990s, when the species as a whole was increasing on the moor. Another specific decline concerns the Postbridge area, where 100-150 pairs were found during 1956 to 1962. They then bred all over the rough ground, even in the lower valleys. They are now confined to the northern higher moorland and the total number of breeding pairs in the mid-1990s was about 20-30 pairs (Dare 1996). The reason for this decline is not known with certainty, but is likely to be habitat linked. With the present policy of reducing stock now well in place, the future may not bode well for the Wheatear.

In 1979 the DNPA/RSPB Survey, a total of 1182 pairs were recorded (Mudge *et al* 1981). A follow up survey of sample areas, covering an area that had 98 pairs in 1979, found 161 pairs in the 52 square kilometre plot, giving about three pairs per square kilometre. The 1992 RSPB Survey of the sample area on the north moor, covered in 1986 and 1979, but with a rather different survey method, gave a total of 316 in the 52 square kilometre plot, about six pairs per square kilometre (Chown 1992). The DNPA High Moorland Sample Survey of 1997, however, found six pairs in a 16 square kilometre area, the same as found in 1979 and 1986, thus halving the 1992 figure. Along with the specific site surveys quoted above, it would appear that Dartmoor in 1992 and 1993 had a particularly dense breeding population. The most recent whole moor survey in 2000 recorded a total of 2593+/-569 pairs, but as 76% of the birds found were male, then there was possibly a minimum of 3823 territories. This total makes Dartmoor's Wheatear of national importance, with between 4.7 to 6.9% of the U.K. population (Geary 2000).

The first birds in spring arrive back on the moor about mid-March. There has been one January record, perhaps of a wintering bird, on Wrangaton Golf Course in early January 1978. Other than this the earliest records have been of one quoted by D'Urban and Mathew at Trowlesworthy Warren in 19 February 1868, and two at Lee Moor on 26 February 1996. There have been several first arrivals between 5 and 13 March, but arrivals most years begin about 15 March, with the main arrival from the third week of the month, and through early April. As well as our own breeding population returning, the good feeding on moorland commons also attracts migrants going further north. These migrants can occur in noteworthy parties at times, as with the 27 on Belstone Common on 2 April 1998, and the 20 on Yellowmead Down on 9 April 1999. Sixty birds were present at Haytor a little later on 20 April 1992. Many of the late April and early May birds are probably migrants of the large and bright Greenland race *O.o. leucorhoa,* although specific attributions are few.

After the young have fledged family parties are present around the moor, and as migration time approaches larger groups are sometimes encountered. Eighty, at Belstone on 27 July 1991 was the largest number noted together at this period. There are many records of about 20 birds together as at Ditsworthy Warren House on 2 August 1998, Stalldown on 27 July 1999, and Kestor 10 August the same year. It is usually noticable that the majority of birds in these parties are juveniles. Breeding birds start to leave the moor from early August, and most are gone by early September. An unusual record of visible migration involved 40 birds flying south and east at Brentor on 11 September 1936. From this time on it is impossible to tell local stock from migrants passing through, although the bulk of September records are probably migrants. There is usually still a trickle of birds passing through in early October, and many of these could be *leocorhoa* race, that tend to move through a little later than *oenanthe.* The only confirmed instances of the Greenland rare *leocorhoa* occurring have been a single in spring at Tavy Cleave on 9 April 1990, at least three out of 60 Wheatears at Haytor on 20 April 1992, and a single male at Routrundle on 2 May 2000, and in autumn one at an unspecified Dartmoor location on 19 August 1960. Most years the last stragglers are seen in mid-to late October, but there have been six records in November, the latest being six birds at Okehampton Camp on the 14 November 2000, and singles at Gutter Tor 17 November 1984, and Prewley Water Works on 20 November 1977.

BLACK-EARED WHEATEAR *O. hispanica*
Very rare vagrant.
A male of the white-throated form was watched at Vitifer on 11 May 1947. This is
the only record.

ROCK THRUSH *Monticola saxatilis*
Very rare vagrant.
On the 25 May 2004 the writer was lucky enough to find a fine summer-plumaged
male Rock Thrush sitting in a small Rowan, in the Erme Valley just upstream from
Piles Copse. It flew over the river and was lost amongst the clitter on the eastern
slopes of Stalldown. Fortunately it was relocated by others later in the day, but could
not be found the following morning. The weather at the time was fine and settled
with high pressure. It was the third record for Devon and the first for Dartmoor.

WHITE'S THRUSH *Zoothera dauma*
Very rare vagrant.
On 11 January 1881 a bird was shot on the edge of Dartmoor. Three or four more of
the same species were also present. D'Urban and Mathew gave the location as Dene
Wood near Ashburton, but presumably this is the Dean Wood near Buckfastleigh as
a Dene Wood cannot be traced.

RING OUZEL *Turdus torquatus*
Migrant breeder and passage migrant. Occasionally a bird will over-winter.
With one noted exception, this has always been a bird of high northern Dartmoor
where it has made its home amongst the rocky tors, clitter, and steep sided coombs.
A link between breeding sites and extensive areas of whortleberry has recently been
confirmed (Jones 1996). It has always been something of a rarity on southern
Dartmoor in the breeding season. In Britain generally it is known that the species had
a steady decline throughout the twentieth century as a breeding bird. Although this
decline was well documented in certain areas, it was not until 1979 that any estimate
was made of breeding numbers on Dartmoor. Twenty nine pairs were then found
during the RSPB/DNPA whole moor survey (Mudge *et al* 1981). A follow up sample
survey in 1986 found 18 territories where 17 had been located in 1979 (Sitters 1988),
confirming little change. Fieldwork for the national *Atlas of Breeding Birds* 1988-1991
found birds in only five 10 kilometre squares, although they had been present in 10
squares for the earlier Atlas in 1968-1972. This suggested a decline, and the DNPA
conducted a survey in 1997 and 1998 to establish the number of pairs on the moor.
The result of this survey was that 29 to 40 pairs were present. Thus the figure was very
similar to 1979, and showed no decline.

Before 1979 one of the few areas to get good coverage was the Postbridge area, which
included Water Hill, Vitifer, Headland Warren, Grimspound and onto Hameldown.
This was, in the mid-twentieth century, as it is now, the main site for Ring Ouzels on

Dartmoor, and the only regular breeding area away from the northern moor. Here the birds take advantage of the numerous old mining gullies, for their steep sides, with dense heather, make ideal nesting habitat. A decline has been noted here over the years. In 1948 five singing males were encountered in a quarter mile of valley (Ware 1948). Between 1956 and 1967, 10 to 15 pairs nested regularly (Dare and Hamilton 1968). A careful search of one square mile in this general area in May 1959, revealed at least 15 pairs, mostly feeding young, but with three nests still with eggs. By the early 1990s this figure had declined to seven to 10 pairs (Dare 1996), and the 1997/1998 survey found six or seven pairs, although the parameters may not have been quite the same. This species is shy and easily disturbed, and human disturbance could well be a factor in the decreasing numbers in this area. One site where birds return every year but do not breed due to humans is the quarry at Foggintor. As early as 1954 birds were reported as deserting here due to disturbance.

The only other site where more than a single pair breeds regularly is Tavy Cleave. Birds have been noted here from at least 1936, and when a count of territories has taken place it has always remained a constant three. This was so in 1957,1978,1988,1995, and 2002. The only year with an increase was 1998 when four territories were found, but only three pairs bred. Certain areas on the northern edge of the moor have shown marked contractions of range in the last 40 or so years, perhaps because of human pressure. On the Taw three pairs were found between Belstone and Taw Marsh in 1963. By the late 1970s pairs were only being reported from the Taw Marsh area, and

Ring Ouzel

by the late 1990s birds had withdrawn to Steeperton Gorge. Likewise on the West Okement, where up to four pairs could be found in Meldon Valley in the early 1970s, prior to the opening of Meldon Reservoir, but in recent years whenever pairs breed it is further upstream around Black Tor. In the south-east of the moor, a regular site from the 1960s to early 1980s was Haytor Down and quarries. Two pairs bred in 1970 and 1971, but the last breeding occurred in 1981, no doubt another victim of human disturbance. Many of the other high, quiet, and out of the way tors on the northern moor have had, and still have, breeding pairs. Long may they continue.

The earliest spring arrival was a bird at Bennetts Cross on 5 March 1994, although there was a bird in the Erme Valley on 19 February 1949. This could have been a wintering bird or an extremely early migrant. More normal early dates are in mid-March, with arrivals most years being in the last week of the month. The main passage is then through to about mid-April. It is difficult to be sure if early birds are local breeders returning, or off-passage migrants feeding up before continuing north. It is thought that the main British population arrives from the southwest, and then moves inland (Wernham *et al* 2002). There is little in the spring gatherings on Dartmoor to suggest anything other than local breeders arriving and feeding together in small numbers, prior to setting up territories. However, there are two records that are significantly different from the norm. On 29 March 1980, an hour before sunset, a flock of 37 flew into the valley at Vitifer, and settled to roost. This is by far the largest spring flock seen on the moor, and the second largest congregation at any time of year. This number of birds arriving together must have been a migrant flock of probably British breeders moving north. The second spring flock consisted of at least 25 birds at Sandy Ford on 4 and 5 May also in 1980. The date of this record suggests that they were birds of the Fennoscandian population that migrate later than British birds.

Recent survey work has shown that many Dartmoor Ring Ouzels are double-brooded, with the first brood fledging about the end of May or early June, and the second fledging late June or July (Jones 1996). There is then a quiet period when few birds are seen and any post-breeding dispersal is not very evident. In other areas it is recognised that birds are very elusive at this time, and that they probably quietly move into food-rich areas, to take advantage of the *Vaccinium myrtillus* berry crop (Wernham *et al* 2002). This presumably also applies to the Dartmoor population, but as yet it is not confirmed. Adults and juveniles moult in July and August, after which they become more conspicuous, gathering in numbers, sometimes with Mistle Thrushes, to feed on the ripe Rowan berries. The frequency that these gatherings occur appear to be less now than 30 years or so ago. The Meldon area had several flocks in the 1970s and 80s, with about 40 on 15 September 1974 being the largest number recorded on the moor. Another noteworthy gathering was 25 on Rowans at Powder Mills on 4 October 1956. They stripped the trees and the last three birds were noted on 9 October, when all the berries had gone. In more recent years the autumn of 2002 was particularly good for the numbers passing through. At least 70 were seen between 13 September and 4 November, with the biggest flocks being 17 on the West Okement on 26 September and 12 at Cox Tor on 18 October. Birds, migrants from further north, are seen throughout October, and by the time the winter thrushes are arriving at the end of the month, birds of the Fennoscandian populations are appearing with them. Each year small numbers occur into November, most in the first 10 days, but one or two later. Interestingly, the November birds are by no means a recent phenonenon, and the Rev. Thomas Johnes noted them on the moor into November in the 1830s (Bray 1836). The latest ever recorded was a bird at Cawsand Beacon on 25 November 1995. By then there is only a fine line between which are late migrants, and which are more likely to be wintering birds. One at Soussons on 15 December 1985 was with Redwings and Fieldfares, and so presumably would have wintered with them. Others seen in December could well also have been associated with the Baltic thrushes. But birds at the summit of Western Beacon on 17 January 1954, Cramber Tor on 29 January 1981, Tavy Cleave on 8 December 1984, and Great Mis Tor on 2 January 1997 all appeared to be far more in tune with their characteristic high moor breeding haunts. A unique record was that of a male that wintered in a small garden at Grenofen from 24 January to 22 March 1987. It arrived in particularly cold conditions, and fed for most of its stay on *Cotoneaster* berries (Jones 1989).

BLACKBIRD *T. merula*
Resident breeder, and probable passage migrant and winter visitor.
The Blackbird has always been a regular and widespread resident of moor edge sites, and as such has been widely overlooked. Very few comments have been made in the literature regarding its status and numbers on Dartmoor.

The 1937 Report of the Devon Bird Watching and Preservation Society mentions that it was widespread, even frequenting quite open regions, and instancing a bird singing from the top of Hen Tor on 27 May that year. In the Postbridge area during 1956 to 1967 the breeding population was thought to be about 100 to 200 pairs, and increasing through the colonisation of the maturing conifer plantations (Dare and Hamilton 1968). In 1956 a pair was found breeding at Headland Warren, close to Ring Ouzels, and sharing a field with them for feeding. This appears to be the first instance of possible conflict for breeding sites between the two species in the area, a conflict that perhaps places some restriction on the later breeding Ring Ouzel to this day. Follow up survey work in the 1990s gave a similar figure at about 100 to 200 pairs (Dare 1996).

The Common Bird Census plot at Harford in 1989 had a total of 21 territories, an increase of 31% on the previous year. The number rose to 24 in 1991, but then went into a decline culminating in only 10 territories in 1997. The population was then quick to recover and had reached 21 territories again by 2000. Counts at Dunsford Wood Nature Reserve over about the same period were much steadier. The 15 pairs/territories in 1995 was about average for the previous 10 years, and the only surprise was the total reaching 22 pairs in 2000, an all time peak for the site since recording began in 1980. This was obviously a good year for the species, as it co-incided with the rapidly achieved peak at Harford. The Dartmoor Moorland Breeding Bird Survey of 2000 recorded birds in moorland fringe areas, in scrubby habitat and associated with forestry plantations. It was estimated that the total moorland population was 608+/- 305 birds (Geary 2000). Counts for the BTO Breeding Bird Survey at Throwleigh Common has given a maximum of 20 birds in a one kilometre square between 1999 and 2002, and survey work for the Devon Atlas Project in 2002 gave a total of 34 pairs in the Bonehill area tetrad.

After the breeding season birds disperse to other areas to feed. An example of this was at Cadover Bridge on 13 August 1998, when many were found out in the open, with a large number of immatures. Migrants no doubt occur in the autumn, but there are not sufficient records to confirm this. Some large parties have been seen in the winter however, and could consist of winter visitors from elsewhere. Forty were seen around Brimpts Farm on 5 December 2002, and the same number at Fernworthy on 9 December 2002. A marked increase in the number of birds around Welltown in Walkhampton, was noted in the cold weather at the end of December 1996, and a male at Plym Head on 12 February 1984 certainly suggested a non-resident.

[DUSKY THRUSH *T. naumanni*]
Very rare vagrant.

On 29 December 1941 R.L. Winter was watching Redwings and Fieldfares at close range near Chagford. Amongst them he noticed a bird that was obviously different. He took field notes, and a description with a sketch of the head and neck area was duly submitted to the Editors of the Devon Bird Watching and Preservation Society's *Annual Report*. Because of the known rarity of the suspected species, R.L. Winter was diffident in putting a name to it, but the Editors accepted it as a Dusky Thrush, and the sketch and description appeared in the 1941 *Annual Report*. A short note later appeared in *British Birds*, and it was later included in the list of Dartmoor birds submitted to the newly formed Dartmoor National Park Authority (Hurrell 1947). It was not included in Witherby's *Handbook of British Birds* however, and probably because of this was not accepted in the later *The Birds of Devon* (Moore 1969), or subsequently in the *Checklist of Devon Birds* (Rosier 1995).

There is no doubt that the description submitted by the finder would not stand up to British Birds Rarities Committee scrutiny, as it is nowhere near complete enough for modern standards, but there is enough of it to make interesting reading, and to leave a thought in the mind that perhaps this could have been Britain's second or third record of Dusky Thrush. Details of the description are given here for completeness, and so readers can make up their own minds about this controversial record. It should be remembered that in 1941, terminology and topographical descriptions of even the most careful observer were not of today's exceptionally high standards.

The bird was noted as obviously bigger than the accompanying Redwings. The head was dark brown, with a broad white eyestripe; the cheeks buffish-white; the back and tail dark brown. There was a chestnut patch on the wings, and not the flanks, and the legs were brown. The observer was certain the bird was neither Redwing or Fieldfare. Unfortunately no details were given regarding the underparts. The Editors accepted it because the chestnut patch on the wings, and not on the flanks, was considered to make the identification as a Dusky Thrush conclusive.

FIELDFARE *T. pilaris*
Winter visitor, and passage migrant.
In autumn the earliest ever migrant was a bird at Meldon Reservoir feeding on Rowan berries on 15 September 1999. There are then four records for the end of the month with seven at Haytor on 25 September 1951, 12 at Wrangaton on 27 September 1968, four at Buckfastleigh on 28 September 2000, and a single bird at Ugborough Beacon on 29 September 2002. Most years the earliest birds are seen in the first or second weeks of October, but at times they do not appear until the end of the month, or even into early November, when larger influxes occur. Some years these influxes can be sudden and dramatic. Some of the largest numbers seen are of birds passing over, and moving south or west to find productive feeding areas away from the moor. The Okehampton and Meldon areas have seen many of these movements, with 800 passing in an hour over Prewley on 26 October 1971, vast numbers moving west over Okehampton between 12 and 14 October 1973, with similar major movements on 23 and 24 November 1987 and 29 and 30 October 1988. On 9 November 1989 1600 birds passed over Meldon Reservoir in an hour, and the same year, on 14 November, about 1000 moved southeast over Okehampton between 7.30 am and 8.30 am. The most recent movement in this area was of 1000 birds moving south-west over Sourton in thirty minutes on 31 October 1996. Similar, if less dramatic, movements have been seen elsewhere on the moor, with at least 200 in small flocks flying up from the Dart valley on 27 October 1989, passing over Venford Reservoir, and moving on to the southwest over Holne Ridge, and 500 moving south over Week Ford, Huccaby on 30 October 2001.

In this period of autumn activity, flocks will also settle in areas to strip the hawthorns of berries before moving on. There are numerous records of flocks of 100/300 participating in this action over the years from all parts of the moor. A noteworthy recent instance was in the Huccaby area in early November 2001. The first indication of unusual numbers were the 500 moving west on 30 October, mentioned above. On the 2 November unprecedented numbers were in the area from Huccaby towards Sherberton and Dunnabridge, with at least a 1000 estimated to be present, taking advantage of the abundant berry crop. Hundreds were present the next day over the whole area, and good numbers remained until 10 November when the action died away, the hawthorn crop no doubt being exhausted (Hibbert 2001). Many other areas experienced unusually large numbers at this time. Weather conditions had been clear, calm and cold, with the wind going around to the north, and thus providing ideal weather for the movement.

Mid-winter numbers are governed by the existing local weather conditions. After the berries have gone birds will take to grass field areas and the like to feed. In mild conditions flocks will happily feed in suitable habitat on the moor, or around the moor edge. However, any period of severe freezing conditions, makes their invertebrate food instantly unreachable, and they leave at once for milder climes. So the presence or absence of flocks during the winter is subject to prevailing local weather, and at times conditions elsewhere. In 1980 a cold front moved through the area on 24 and 25 December, with snow on the hills on 26 December. A weather movement of several flocks totalling over 800 birds were seen on 24 December, no doubt moving through ahead of the front. A winter roost was known in the plantation at Soussons in the early 1970s, and several hundred birds were watched entering the trees there on 15 December 1973. The improved short sward fields in the West Webburn valley always attracted Fieldfares in winter. About 400 were present there on 17 February 1957, and at least 700 on 7 February 1999.

By late March the wintering birds move away east and north, and migrants are beginning to pass through. A large gathering estimated to be 1000 at Hennock Reservoirs rather earlier than this on 5 March 1988, could have included migrants. Numbers reduce dramatically in early April, but this was also the time of two exceptional records. On the 10 April 1966 about 2000 migrant Fieldfares passed over Lynch Tor moving northeast, and on the 9 and 10 April 1989 large numbers were noted at Yarner Wood flying up the valley towards the open moor. A flock of 200 were seen at Lustleigh on 21 April 1990, and 12 were still present at Beardown on 1 May 1966. On the 23 April 2004 a bird at Shipley Bridge was in song. An amazing flock of 56 was in the West Webburn valley on 8 May 1957. The last spring record was a bird in the Avon valley on 30 May 1960. There is one mid-summer record of a bird seen and heard at Ditsworthy Warren on 29 July 1972.

One phenomenon that should be touched on briefly is ground roosting. This was first noted by observers waiting for Hen Harriers near Birch Tor on winter evenings in December 1996. It was noted again in 1998, and on 17 January 1999 a flock of over 300 came into roost at dusk. It was subsequently observed in areas close by at Warren House Inn, Vitifer and Merripit Hill, but in much smaller numbers. The bird's requirements appear to be just a sufficient height of ground cover, preferably heather, but bracken was also seen to be acceptable. It is not known how widespread this habit is, but the fact that at least 75 birds were noted at dusk on the summit of Yar Tor on 14 November 2001, would appear to suggest that it could be more common than at first anticipated.

SONG THRUSH *T. philomelos*
Resident breeder, passage migrant and probable winter visitor.
The Song Thrush has always been scarcer on Dartmoor than the Blackbird, and during the 1990s when the nationwide plight of the Song Thrush became evident, its numbers on the moor, as elsewhere, were at an all time low.

Historically there are very few records for comparison, but in the Postbridge area 50 to 75 pairs were estimated to be present in the 1950s and 60s. They were very badly hit by the 1962/63 winter, with losses up to 75%, but had regained their former numbers by 1966. They increased as the maturing conifer plantations were colonised (Dare and Hamilton 1968). By the early 1990s the population was found to be drastically reduced with only 15 to 25 pairs present, and most of these in the plantations (Dare 1996). By the end of the 1990s there was the general feeling that numbers were beginning to increase nationally. This was reflected in certain casual records in our area, although the rise at first was rather slow. As the early years of the twenty-first century progress, it does appear that this species is becoming commoner again. At the Harford Common Bird Census area, numbers were at a low level throughout the 1990s, but there was a difference from year to year. Only a single territory was occupied in 1990, but four were occupied in 1991. Throughout the rest of the decade two or three pairs were present, except for 1999 when the total rose to five. At Dunsford Wood Nature Reserve the average was 4.1 pairs in the same period. This increased to eight pairs in 2000, but had dropped back to six in 2002. The improving fortunes of the Song Thrush were shown by survey work for the Devon Atlas Project in 2002. Some very good counts of singing males/pairs were obtained, including 12 at Huccaby, 10 in Holne Chase, and four each at Yennadon Down, the West Dart valley north of Hexworthy, Bonehill and Bench Tor.

In the autumn when the Redwing and Fieldfare arrive, or possibly just before, migrant Song Thrushes pass through the area. It is probable that most of these migrants are of the greyer continental race *T.p. philomelus,* but the plumage differences are subtle, and as these migrants are always very shy, few are ever racially identified. In October 1998 there was a very large influx of migrants from Scandinavia onto the East Coast, mainly between 1 and 7 October. The movement was noted in the West Country, with 120 at Portland Bill on 7 October, and numbers over Plymouth and Berry Head the same day. What was unusual was that this influx was noted at certain Dartmoor localities. In the Deancombe valley at Burrator, at least 30 were feeding on Rowan berries with Ring Ouzels and Mistle Thrushes on 5 October, with 12 still present on 6 October. On 8 October at least a 100 were found in Lee Moor Woods, with another 20/30 nearby around Cadover Bridge. It is probable that autumn passage birds occur on or over the moor most years, but these numbers were exceptional. It is possible that some migrants may stay to winter with Fieldfare and Redwing, but records are lacking. There has never been any indication of a northward or eastward passage over the moor in spring.

REDWING *T. iliacus*
Winter visitor, and passage migrant.
The first autumn Redwings appear usually a little before Fieldfares, during the first week of October. There have been five September records, the earliest being 10 or 12 at Thornworthy on 13 September 2004. Three were at Wrangaton on 22 September

1962, and the other September records, all on 27 September, were one at Bellever in 1970, three at Two Bridges in 1974, and 12 at Fernworthy in 1999.

It is normally the middle of October before any large numbers arrive. On 13 October 1972 many hundreds passed over Prewley, and on the same date in 1980 there was a large early morning movement over Okehampton and several hundred moving east over Dousland. These waves of migrants, no doubt weather related, then occur through to early November. Large movements are not observed every year, but when seen they can be quite spectacular. At Gidleigh on 19 October 1986, over 1000 flew over during the day, and thousands over flew Okehampton on 6 November 1990. The area around Fernworthy Reservoir and plantations has attracted very large flocks in autumn over the years. Hundreds were watched dropping down to feed on 19 October 1989, and then moving on south, 900 were present on 24 October 1992, and a totally unprecedented 4000 moved over the area on 22 October 1995. The last coincided with a large passage on the south Devon coast the same day. A large mixed flock containing about 1500 was seen at Brisworthy on 16 November 2002. This was interesting because the autumn was not good for the species generally, with no other large flocks reported, and a general shortage of birds. When the autumn flocks are migrating, there is evidence to suggest that at least some of them may be moving somewhat at random in their search for food. An instance of this was in 1983 when at least 250 were reported moving northeast over Grenofen on 21 October. The same day 240 moved east over nearby Dousland, and the next day at least 400 moved south over South Zeal.

As with their companions the Fieldfares, Redwings tend to stay in areas in late autumn and early winter until the berry crop has been exhausted. They then move to any suitable grassland habitat, or if none is available, move out of the area. Winter flocks are very much weather governed, and are not as large as the autumn gatherings. Birds leave an area at the onset of low temperatures, and if anything, they are away before Fieldfare flocks. When birds are forced to leave areas, the visible migration can be spectacular. On the 4 January 1955 flocks were watched moving south over Ugborough Beacon after snowfall. A thousand passed over in 11 minutes, and the passage continued for at least an hour. The maximum regular winter flocks in the past tended to peak at about 400/500, but unusually large numbers occurred at Lee Moor on 12 December 1978 when 1000 were present, and at Fernworthy on 20 January 1989 when 987 were counted. Also in 1989 at least 600 were at Okehampton on 25 January. In recent years wintering flocks have been smaller, peaking at about 200/300. An interesting record occurred in the severe weather of January 1963 when a dead bird, picked up at Bittaford, was found to have been ringed as an adult on migration on 19 September 1961 at Tampere, Finland.

By early March flocks diminish, and birds are heard at night passing overhead on return migration. Flocks will come off passage at times to feed and roost, as with the 100 at Yarner Wood on 4 March 1990, and 11 March 1992. There are still a few around most years in late March, but April records are very few. There are two records of spring passage flocks in April. On 10 April 1966 at Lynch Tor 100 were watched drifting northeast with a far greater number of Fieldfares, and on 15 April 1980 40 flew southeast over Princetown. This record is the latest in April. There has been one totally exceptional May record, when a bird was seen at Haytor on 28 May 1998.

Ground roosting has not been recorded on Dartmoor as it has for Fieldfares, although a few birds have been flushed at Merripit Hill in near darkness in recent years, suggesting that it may occur, and at Huccaby Farm at least 20 roosted in gorse during October 2001.

MISTLE THRUSH *T. viscivorus*
Resident breeder. Also probable passage migrant. Winter visitor status uncertain.
The song of the Mistle Thrush is one of the first to be heard, with many males singing in December or even late November. The population on Dartmoor would appear to be stable, with good numbers of birds inhabiting the conifer plantations, and moor edge broad-leaved woodland.

The population in the Postbridge area between 1956 and 1967 was estimated to be 10 to 20 pairs. They suffered a reduction of about 25% in the 1962/63 winter, but numbers were fully restored by 1967 (Dare and Hamilton 1968). In the early 1990s about 15 pairs were found, and the population was thought to be still increasing in the matured conifer plantations (Dare 1996). At the Harford Common Bird Census site, three territories in 1989 had risen to six by 1991, and this level was maintained until 1995 when the occupied territories dramatically dropped to two. There was an increase to three in 1996 to 1998, then four in 1999 and five in 2000. There are few other records from constant effort sites, although two or three pairs were typical at Dunsford Wood Nature Reserve from 1995 to 2002, and two pairs were present in Yarner Wood 1990 to 1994. Casual counts elsewhere have included 10 pairs in the Burrator area in 1996, three pairs at Bellever in 1997, and three pairs at Huccaby in 2001. The fieldwork for the Devon Atlas Project in 2002 provided more good numbers in the Dartmoor area, with maximum numbers being 16 pairs at Holne Chase, four pairs in Huccaby Cleave and between Deancombe and Norsworthy Bridge, Burrator, and three pairs in North Park Wood, Holne.

By mid-June, family parties, feeding on short sward grassland on the moor and in moor edge fields, are beginning to form loose flocks. They move away from breeding territories, and become scarce for a time in certain areas. At Huccaby, where they are fairly common residents, there were no records between 18 June and 18 August 2000. Presumably residents had departed to join the wandering flocks feeding on the open moor. From August on, these flocks start to take advantage of the ripening berry crop. Flocks of 20 to 30 are commonplace most years, but occasionally they will join together to form a much larger gatherings. A loose flock of 80 was present at Headland Warren on 25 July 1990, and on 9 August 1956 about 100 in a large flock were seen at Postbridge. 2000 was a good breeding year on the moor, and large numbers of juveniles were noted in July in the Thornworthy area. At Huccaby, where there had been a mid-summer absence, numbers began to increase from late August. Double figures were counted on many days until early September, but there was nothing to predict the influx of 7 September. On that day Richard Hibbert watched birds in flight, streaming in a northerly direction from the lower O Brook/Coombestone slopes towards Clay Brake and the rest of the Huccaby peninsula. In all there were at least 150 birds, making this the largest flock ever in Devon. Whether this number included birds from elsewhere, or whether it was a gathering of all the flocks in the area, coming together to take advantage of the exceptionally good Rowan crop, is uncertain. Numbers were present for the rest of September and early October, but the maximum was only 26 until the period of 13 to 16 October when about 50 were present (Hibbert 2000). Flocks of 40 were also noted that autumn at Scarey Tor on 23 September and Sticklepath on 4 October.

An increase in October numbers could well indicate migrants from elsewhere moving through the area. These birds, sometimes noted through to mid-November, are often associated with Ring Ouzels and winter thrushes. It is possible that they are from the Scottish breeding population, known to be migratory, or even from the continent. The biggest flock at this time of year was at least 70 at Burrator on 15 November 1986. It is thought that at least some first winter birds, bred in southern Britain, migrate south late in the autumn, thus perhaps accounting for the general scarcity of birds from late November through the winter. Adults tend to stay in, or close by, breeding areas, and defend winter feeding territories, breaking into song when the need takes them.

GRASSHOPPER WARBLER *Locustella naevia*
Migrant breeder and passage migrant.

This is a species that is very difficult to locate. If it was not for the bird's reeling song, given from cover, it could well be thought almost non-existent on the moor. However, careful listening in damp areas with rank vegetation and willows will at times reveal its strange metallic song, although the bird usually remains hidden. It can be imagined that the relatively few casual records of birds reported go no great way to estimating the total population for the moor.

The species has a habit of being quick to colonise newly available areas, only to depart just as rapidly after the habitat changes. This happened in the young conifer plantation at Soussons between 1956 and 1962. Birds readily accepted the long grass and herbage cover between the trees, and in a few years a population of 10-25 males was present there and in nearby bogs. Once the trees passed their scrub stage, by about 1967, the plantation sites began to be unsuitable. By the early 1990s only two to five territories could be found in the area, in the valley bogs of the West Webburn Valley, and around Postbridge (Dare 1996). Today this figure could be higher due to a few birds moving into dense heather habitat around Vitifer and Bennetts Cross, and clear-felling of the plantation making certain parts suitable again. In the spring of 2004 up to seven reeling birds could be heard in the Warren House/Vitifer area, with an additional three around Powder Mills.

D'Urban and Mathew mention that in the late nineteenth century Lord Lilford found six nests on a small patch of ground on the northern slopes of Dartmoor. This could well refer to the area of small marshes that exist on the northwest edge of the moor. These have been productive for the species over the years. In 1949 and 1950 up to three pairs bred on marshy ground at Sourton, and between 1975 and 1979 up to four reeling males were present at Prewley Marsh. The area around Fernworthy Reservoir has held birds from as far back as 1948, with four reeling there on 30 June 1951. Birds still turn up in the same area today. The sewage works area by Oakery Bridge on the Blackabrook had records during the 1990s, with a maximum of three birds reeling. Few of the professional surveys in recent years have really got to grips with the species, but the Dartmoor Moorland Breeding Bird Survey of 2000 looked at larger areas of suitable habitat, and was able to come up with an estimated total for the whole moor. Areas of Rhôs pasture, largely along the north-east and east edges of the moor were surveyed, and found to be the optimum habitat for the species on Dartmoor. This habitat, characterised by rank *Molinia/Juncus* and scrub and subject to controlled grazing regimes was estimated to hold a total of 43.90 +/- 13.80 males/territories. Valley mire habitat accounted for 14.64+/-10.53 males/territories, and open moor sites were estimated to hold 18.46+/-9.23 males/territories. The overall population for the moor was thus estimated at 77+/-34 males/territories (Geary 2000).

The first birds usually arrive in the last week of April, although occasionally earlier. In 1966 first arrivals were particularly early with birds at Tavistock on 15 April, and Wrangaton on 16 April. Males will at first move into quite unsuitable territories at times, before moving on. Two birds reeling in bracken and gorse on Holne Moor on 22 April 1998, was an instance of this. Another bird in rather unusual habitat was reeling in rushes on the East Dart at Sandy Hole Pass on 30 July 1995. The late date perhaps indicated it was a non-breeder or early returning migrant. Birds have an extended song period. Song well into July is not unusual, although one still vocal on 2 August 2002 at Cold East Cross, was certainly late. With this secretive species, as soon as the song stops, the records usually stop as well. A few birds are noted on autumn passage, and these have a knack of turning up in atypical habitat. September migrants have included birds on Okement Hill on 8 September 2002, near Eastern Whitebarrow on 10 September 2001, and on Cosdon Beacon on 13 September 2002. A *Locostella* species, probably a Grasshopper Warbler, was seen on Walkhampton Common on 2 October 1996, and the latest bird in autumn was an individual at Kennick Reservoir on 14 October 1995.

SEDGE WARBLER *Acrocephalus schoenobaenus*
Migrant breeder and passage migrant.

The breeding population of the Sedge Warbler on Dartmoor is not large. The majority of breeders occur in the Rhôs pasture areas, mainly on the east and northeast fringes of the moor. This is an area that is not monitored regularly, but when members of the Dartmoor Moorland Breeding Bird Survey surveyed this area in detail in 2000, they found 10 territories in five systems, and by extrapolation estimated the total population for the Rhôs pasture areas to be 55.81+/-28.05 territories. The territories were found particularly in wetter areas of rank *Molinia* with scrub (Geary 2000).

Away from the Rhôs pasture, pairs are found regularly in only a few tradional sites offering a similar type of habitat. The main site has always been on the River Plym around Cadover Bridge. Birds have bred there in overgrown clay-pits since at least 1969, although their presence was noted since 1962. Numbers have varied over the years, but have never exceeded three pairs. In the 1950s and 1960s birds were noted in the Postbridge area survey, usually along the West Webburn River, Broadaford Brook, Walla Brook and Stannon Brook. Their status was thought to be uncertain, and they were taken to be non-breeding summer visitors (Dare and Hamilton 1968). However, in recent years a small number bred each year (S. Needham *pers. comm.*1992), and the 2000 survey found six territories in the general area. Another area that attracts birds is the section of the Blackabrook River near Oakery Bridge. One or two singing males have been noted there nearly every year from 1990, and in 1997 a male was located just downstream at Batchelor's Hall Bridge. Further downstream again, after the Blackabrook has joined the West Dart, three singing males were found on 4 July 2000 at Princehall Farm. In recent years others have been found at Fernworthy, Swincombe valley, Willsworthy, Wedlake Farm, and Muddilake and Criptor Newtakes.

With so few breeding it is difficult to say just when they arrive. One on the Plym at Cadover Bridge on 1 April 1999, was very early, and it is likely that most of the local breeders arrive towards the end of April. On occasion birds take up temporary territory in unsuitable locations, as the one singing from a village garden hedge at Horrabridge on 11 May 2001. On autumn passage they can occur in all types of medium-height herbage. It is not too uncommon to find them in bracken either, and on 21 August 1938 a migrant was found in a thorn bush on a dry bracken-covered slope at Ugborough Beacon. This, incidentally would appear to be the latest date a bird has been seen on Dartmoor.

MARSH WARBLER *A. palustris*
One breeding record.
In 1954 the naturalist F. Howard Lancum found a pair of breeding Marsh Warblers in South Devon. They were successful in raising young, and were watched feeding the young on two occasions. The locality was not made public at the time, but later it was shown to be Burrator Reservoir (Moore 1969).

Little detail was ever made known, but it was obviously fully accepted by the county authorities at the time. At first glance this looks to be an almost unbelievable record of a rare British bird breeding well outside its normal very restricted range, but further research shows that the record is far from impossible. Firstly, there is habitat around the inlet end of the reservoir with medium to tall vegetation, willow scrub, etc., that could be acceptable for the species. The mid-1950s was at the end of a period of some expansion, and the begining of a contraction of range and decline in numbers in the late twentieth century. Most important, however, is the known sporadic nature of breeding in the species. The West Midlands population, before its more recent demise, was noteworthy for its continuity. Elsewhere most records were of transient breeding, with birds occupying a site successfully one year, but never returning again (Gibbons *et al* 1993). Birds have bred as far away from their normal range as Orkney, so a one-off breeding on Dartmoor is not too unlikely.

DARTFORD WARBLER *Sylvia undata*
Resident breeder.

The rise of the Dartford Warbler as a breeding bird on Dartmoor has been the ornithological success story of the late 1990s. Up until 1992 records had been very few and far between. D'Urban and Mathew refer to birds being seen on Dartmoor and near Ashburton, but give no details. The *Devon Bird Reports* for 1949 and 1950 show singles in the south and southeast areas of the moor, but again give no details. On the 1 May 1952 a male was seen at Fernworthy, and on 28 May 1959 a bird was recorded on Whitchurch Down. Both these dates were interesting in as much as they could have indicated breeding.

There were no records then until one was watched on Trendlebere Down on 12-15 December 1992. The late 1980s had seen a big increase in breeding pairs on the East Devon Commons, their traditional stronghold in the county. The dispersal from there first showed on the coast, but by the autumn of 1988 a pair had taken up residence on Haldon, and in the autumn of the following year a bird had moved westward, being seen on Knighton Common on 14 November. The Trendlebere bird was obviously a part of an ever-westward range expansion. In the next few years numbers built up at Trendlebere and breeding occurred. The colonisation was so rapid that by 1996, when a disastrous fire swept the Down, English Nature was able to report to the press that the number of Dartford Warbler territories destroyed had run into double figures. By 1994 birds had spread to other areas close by, with reports coming from Buckland Common in November, and the moor edge site of Ramshorn Down in October. Breeding was proved at Buckland Common in the May of 1995, and probably occurred at Rushlade Common. In the autumn of 1995 a bird was seen a little way to the north at Bowerman's Nose. The first sign of further westward expansion was noted in 1994, when a male was found in the Dart valley near Venford Reservoir in June.

The fire on Trendlebere Down in 1996 ruined the main Dartmoor site, but no doubt birds moved out ahead of the flames, and perhaps by coincidence or otherwise 1996 was noted for further range expansion. Birds were seen at Buckland Beacon and Blackslade Down in July, but more importantly they began turning up at sites on the western edge of the moor. One snowy February day one surprised an observer in scrubby low western gorse near Routrundle in the Walkham valley. Later that spring birds were reported from nearby Criptor Newtake and Yennadon Down. At

171

the extreme southwest corner of the moor a pair bred at Roborough Down, and on the northeast edge a bird was seen at Mardon Down, near Moretonhampstead in September. The consolidation continued into 1997, with more suitable habitat being taken up in the southeast, and a breeding pair observed at Henlake Down, Ivybridge in May. Another new site in the southwest was Shaugh Moor, where a bird was seen in mid-May. In September of 1997 a female was watched in Belstone Cleave, a previously unrecorded area.

Between 1998 and 2004 the expansion continued, with new sites turning up birds every year, and breeding pairs increasing at previously established sites. On the eastern edge, Cosdon Beacon, Throwleigh Common, Round Pound near Kestor, and Cranbrook Down recorded birds. Individuals were found in the Bovey valley at Harton Chest and Hunters Tor, and in the area of the two Darts the species was noted near Aish Tor, Holne Moor, Yar Tor and Yar Tor Common, Huccaby Tor, near Laughter Tor, and on the southern edge of Riddon Ridge. By the River Avon, Dockwell Ridge, Smallbrook Plains, Avon Dam and a site near Brockhill Ford had records, as had the nearby Harbourne Head area. Crownhill Downs had several breeding pairs, and also in the southwest Wigford Down produced records. The west side of the moor had reports from Pew Tor and West Down, actually just outside the western boundary of the National Park. Only the western fringe of the moor, between Tavistock and Okehampton appeared to be largely unaffected by the range expansion. Even here a pair was watched on the lower slopes of Brat Tor in February 2001, and a bird was seen near Great Nodden in the summer of 2004.

Almost without exception, sightings of birds have been in habitat dominated by gorse, with the low western gorse *Ulex gallii* being especially favoured. However, on two occasions in the central portion of the moor, birds have been seen in areas of

Dartford Warbler

tall vigorous heather. The first was a male singing on the slopes of Headland Warren on 17 May 1999, and the second a female/juvenile feeding in deep heather on the western slopes of Water Hill, near Statts Bridge on 26 October 2003.

These records, together with a bird feeding nearby in gorse on Stannon Hill on 25 November 2002, are almost the highest for birds on Dartmoor, all between 410 and 420 metres above sea level. The record for the highest, however, must go to a bird seen on Hew Down, near Watern Tor, on 17 April 2004 at about 500 metres, and surely too vulnerably high into wild uplands for a heathland species. It is also no doubt the highest recorded in Britain.

There have now been records from over 50 sites on Dartmoor. Many of the sites have been noted with birds present in only one or two years, but as casual recording does not allow for an annual census of sites, it is quite possible that the majority of sites are occupied each year. The Dartmoor Moorland Breeding Bird Survey of 2000 found a total of 14 singing males, and through extrapolation it was estimated that a total population of 65+/-32 could be present (Geary 2000). This would appear to be entirely reasonable.

One point that should be mentioned, and in fact could be of assistance to anyone trying to establish the presence of Dartford Warblers in an area, is their close association with Stonechats. This has been noted on several occasions in the literature, and has been observed many times on Dartmoor. It applies to birds in the breeding season, as well as wandering individuals in the autumn and winter. The alert nature of the Stonechat, with its constant scolding if possible danger is present, appears to be used by Dartford Warblers as a warning system. A scolding Stonechat will often bring any nearby Dartford Warbler to the top of the gorse to investigate. This can often prove very useful to an observer, who otherwise may not be aware of the warbler's presence. There must be a certain comfort factor in the nearness of a Stonechat, as when near Statts Bridge on 26 October 2003, a pair flew off whilst the Dartford Warbler known to be accompanying them was deep in the heather. The warbler eventually came to the heather tops and finding the chats gone, became very agitated, calling constantly until relocating them some 200 metres away across a small stream and valley, it flew directly to them.

SPECTACLED WARBLER *S. conspicillata*
Very rare vagrant.
In the afternoon of 3 June 1999 the writer was searching gorse on Roborough Down for Dartford Warblers. Whilst working through a patch, a loud burst of song was heard from close by. The singer was hidden, but the obviously *Sylvia* - type song was not recognised. The bird then showed itself briefly, as it shot across between gorse bushes. The initial impression was of a Whitethroat, but the song was wrong for that species. It was seen fleetingly several times in the next 45 minutes or so, and the long rolling scold given by the bird added to the certainty that it was not a Whitethroat.

After research that evening and discussion with a colleague, a male Spectacled

Warbler did appear to be a real possibility. The bird was seen again on 4 June, and field characters, plumage and song all matched Spectacled Warbler, incredible though it seemed. The identification was confirmed by others on the morning of 5 June, and the bird was watched by hundreds during that day and on 6 June.

It would appear quite possible that the bird had been present for some time previously. Although not specifically identified, a Whitethroat-type bird with a strange song had been noted in very much the same area of the Down in mid-May (June Smalley *pers. comm.*1999). It was the first record of this Mediterannean species for Dartmoor and Devon, and only the third for Britain.

The locating of this bird proved something of a lesson. Dartmoor is known for its specialist species, but not for rarities. A singing Spectacled Warbler on a well-used moor edge down, showed that the totally unexpected can happen, and our eyes and minds should not be closed to this fact.

LESSER WHITETHROAT *S. curruca*
Rare passage migrant. Has bred.
The wildness of Dartmoor is not well-suited to this lover of bushy thickets and brambles. However, it has bred on at least three occasions, at Ashburton in 1960 and 1961, and at Lustleigh in 1967. Other than the proved breeding, all other records have been of singing males, as would be expected. Some of these could have culminated in breeding attempts, but given the scarcity of the species, most were probably birds taking up temporary territory and then moving on, when no mates were found. One or two very obvious migrants were also noted singing in atypical habitat.

The area around Buckfast has attracted more birds than any other locality. Singing males were noted there in June 1952, May 1953 and 1963, and April 1994. A bird was at nearby Buckfastleigh in May 2001. At a rather more exposed location, a bird was singing at Shaugh Prior in May 1949. Also at a rather unexpected area was a male at Burrator in June 1981. In more suitable sheltered areas individuals were seen at Neadon Cleave, Manaton in May 2001, and at Bridford in June 2002. Trendlebere Down was an unusual location for one in July 1987, but a migrant singing early one morning from heather near Bennetts Cross on 14 May 2003 was totally unprecedented.

WHITETHROAT *S. communis*
Migrant breeder, and passage migrant.
Very little was recorded about the Whitethroat numbers prior to the 1960s. It was a common migrant breeder in suitable habitat almost everywhere, whether on the coast or inland. Then in 1969 came the first effects of the disastrous Sahel droughts. Over the next few years these droughts, in the species's African winter quarters, caused a massive mortality. Nationally, Whitethroats went from a common species to a rare one almost overnight, and it took many years for the former levels to become re-established. On Dartmoor this recovery never fully took place.

In 1937 there was comment in the *Devon Bird Report* that the species was very local at higher altitudes of Dartmoor, although it was found at Brimpts Wood at over 1000 feet. By the mid-1950s however, probably helped by the newly formed scrub habitat in the conifer plantations, records from cultivated land suggest that a marked increase had taken place, with 100-150 pairs breeding in the Postbridge area alone (Dare and Hamilton 1968). Coverage of the same survey area during 1990-95 produced only one singing male (Dare 1996).

Elsewhere on the moor, although no direct comparisons with earlier years are possible, the species now has a similarly reduced and patchy distribution. Odd pairs can turn up in any likely gorse or scrub area, but there does appear to be a slightly stronger population on the drier eastern side of the moor. The Dartmoor Moorland Breeding Bird Survey of 2000 found the Whitethroat scarce on the moorland periphery, but did locate six singing males on Easdon Down. Three were found in the Shilly Pool area of Throwlcigh Common in 1996, with singles in several years since. Small isolated moorland areas on the eastern boundary have also proved productive, with three pairs at Cranbrook Down in 1997, and three pairs at Mardon Down the same year. Observers being confined to roads in the foot and mouth outbreak of 2001 turned up breeding pairs that might well have been otherwise missed. In late May seven pairs or singing males were found along the road from Heatree Cross to Cockingford Bridge. In 2002 survey work for the Devon Atlas Project confirmed six pairs in the Bonehill area tetrad, and three at Blackslade Mire. The Harford Common Bird Census site held one pair each year from 1988 to 1998, and Fernworthy has had a pair most years from at least 1997, with breeding proved in 1999.

Birds usually arrive from the last week of April, with one at Buckfast on 19 April 1957 being the earliest. After the breeding season, birds will disperse in an area prior to migration. They can then be found in habitat not used for breeding, particularly areas of bracken. Dispersing family parties or migrants are seen throughout August, with odd birds into September. The latest recorded was a bird at Fernworthy on 25 September 2002. An interesting occurrence, probably involving a family party on migration, or at least moving away from the natal area, was noted on 16 August 2002, when several birds were seen on the north slope of the O Brook, above Saddle Bridge.

GARDEN WARBLER *S. borin*
Migrant breeder, and passage migrant.

This plain warbler with a beautiful song is to be found on Dartmoor in woodland with a well-developed understorey, and also in scrub areas away from mature trees. It is particularly suited to young conifer plantations in their scrub stage. It has also been shown to inhabit garden shrubberies, blackthorn thickets and willow bogs (Dare 1996). Although less common than its close relative, the Blackcap, it inhabits more high ground areas around the edges of the moor, thus reflecting the species's requirement for scrub cover, as against the Blackcaps liking for more open canopy.

Although evidence is not substantial, there appears to be a certain instability in the moorland population from year to year. During survey work in the Postbridge area in 1990 nine singing males were found, but only three in 1991 and none in 1992. The regular number of pairs in the region was estimated to be about five to 10, so there are obviously good and bad years. This was also shown in the Common Bird Census site at Harford, where between 1987 and 2000 the number of territories occupied yearly varied between none in 1993, 1995 and 1996, to five in 1987 and 1998, and a peak of six in 1992. Four singing males in Grenofen Woods in 1995, were the first noted in the area for seven years, and the fluctuation in numbers was noted at Yarner Wood in 1980, when they appeared unusually numerous. The causes of these fluctuations are not known, as the species has been doing well nationally since about 1990. It is possible that problems in their African wintering grounds may be a factor, or perhaps competition at some marginal habitat sites with Blackcaps that arrive earlier and are intolerant of Garden Warblers (Gibbons *et al* 1993). In other areas birds have been more regular with three or four territories at Burrator, and the same number at Fernworthy in recent years. Exceptionally good numbers were located in the area of the Dart valley during Devon Atlas Project fieldwork in 2002. Between Dartmeet and Holne Bridge at least 15 singing males were found, mainly on the edge of valley woods, and in the Hexworthy area a further nine were found in the West Dart valley and the farmland to the north of the village. Three to five territories have been regular at Huccaby in recent years.

The first birds normally arrive in the last week of April through into early May. Early birds in mid-April are not too unusual, but particularly early individuals were at Harford on 1 April 1999, and Shaugh Bridge on 9 April 1982. After the breeding season, adults and young depart for the south in late August and September. Any post-breeding dispersal is poorly recorded on Dartmoor, and there are few records after the young have fledged. The latest birds noted were two at Holwell Tor on 20 August 1986. Thus, although its status as a passage migrant is probably correct, it is stated rather tentatively as proof is lacking.

BLACKCAP *S. atricapilla*
Migrant breeder, and passage migrant. A few winter.
The Blackcap gets little mention in early literature regarding its status on Dartmoor. The early *Devon Bird Reports* only mention the few wintering records, mainly in the towns on the moor edge, and it is not until the Postbridge survey work of the 1950s and 1960s that any depth is given to the records. Here five to 10 pairs were present in Bellever Plantation, and singing was first noted in Soussons Plantation in 1960. One or two pairs were breeding in gardens around Postbridge, and in deciduous copses in the lower West Webburn Valley. The yearly total up to 1967 for the survey area was thought to be five to 15 pairs (Dare and Hamilton 1968). By the early 1990s the total had increased to 20-25 pairs, with at least 13 in conifers (Dare 1996).

The Blackcap is a very successful species in Britain, and is still increasing. It overlaps in favoured habitat with the Garden Warbler, but is more a bird of the tree canopy in deciduous woodlands, and lacks the Garden Warbler's acceptance of thicket scrub. On Dartmoor, suitable sites, both deciduous woodland and conifer plantation, hold a stable and probably increasing population. Several wooded valley sites hold good numbers, as on the Teign, where six were singing at Steps Bridge on 24 May 1990, and three at Fingle Bridge on 18 May 2001. Recently the Dart valley woodlands have been surveyed for the Devon Atlas Project, and the numbers recorded have included 13 pairs of singing males in Holne Woods, seven in Holne Chase, four below Bench Tor, four in the valley north of Hexworthy village, three downstream from Dartmeet, and a further 10 at least at other sites. Fernworthy woods have held up to five pairs since 1997, the same number as noted here in 1966. Woodland around Burrator usually has about three pairs. A significant increase to 11 was noted in 1998, but not since. In the east of the region the well-watched Dunsford Wood Nature Reserve recorded a steady average of 5.4 pairs yearly from 1980 to 1995. In the southwest, Dartmoor's only Common Bird Census site at Harford had a steady one or two pairs from 1987 to 1999, but then an increase to four pairs in 2000.

In recent years the first singing birds have been heard in late March, with a bird at Horrabridge on 28 March 2001 being the earliest. However, with the increase in wintering birds, it is possible that these first songsters may have been about all winter. If this is the case, then the singers are not of the British breeding stock, as our wintering birds are considered to be from central European populations. The bulk of the population arrives in the latter part of April and May. After the breeding season local birds are around until August, when they quietly move south. Migrants from the continent, and perhaps northern British breeding birds, are then beginning to show on their extended migration period, which lasts through until November. Migrants are not too uncommon in September, although five at Fernworthy on 25 September 2002 was an unusually high number, and a male at Huccaby on 30 October 2001 was very late for a central Dartmoor location.

Individuals seen in November may still be migrants passing through, or they may be birds arriving to winter. They are almost invariably seen in urban locations, where the availability of food no doubt makes for an easier life during the cold months. Okehampton, Ivybridge, Tavistock and more recently Buckfastleigh, have all had their share of wintering birds. Buckfastleigh has proved particularly well-blessed with three or four regularly most winters from 1996. Most birds depart from winter sites in late February or early March. More rural locations in the winter, where birds are much rarer, have included Crapstone in 1953, Yelverton in 1963, 2001 and 2002, with two males and a female present the last year, Cornwood in 1974, Grenofen and Horrabridge in 1982, and Liverton in 1996. A record, especially interesting because of its high Dartmoor location, was of a female coming to a bird table at Bellever from 15 December to the end of the year 1968.

YELLOW-BROWED WARBLER *Phylloscopus inornatus*
Very rare vagrant.
There are only three records, which may in fact have only been two individuals, of this vagrant from the east. On 23 December 1982 one was found at Okehampton. This was an atypical winter date and no doubt referred to a bird that had arrived in the country the previous autumn. Two weeks later on 4 January 1983 a bird was discovered at South Zeal, only about four miles away. This could have been a different bird, but given the great scarcity of the species in Devon, plus the closeness to the Okehampton site, it was probably the same. The only other record is of one that was found on a typical autumn date of 5 October, at Lee Moor clay works in 1997.

WOOD WARBLER *P. sibilatrix*
Migrant breeder and possible passage migrant.
The Wood Warbler in Britain has always had its stronghold in the west of the country, where it has been dependent on closed-canopy woods with little understorey and ground vegetation. On Dartmoor, woods of this type are not too widespread, but they do occur, especially on the drier eastern side of the moor, and in the Dart valley.

Although now 20 years old, the most useful source of information on Wood Warblers comes from the BTO species survey results of 1984/85 (Goodfellow 1986). In this survey the most heavily populated 10 kilometer square in Devon was SX77, which included the hot-spots of Yarner Wood and much of the wooded areas of the Dart and West Webburn rivers. The total number of singing males in this square was 132, giving a very high average of 8.8 in each of the 15 tetrads that recorded birds. Several of these tetrads, though, held exceptional totals, and are worth recording for any future comparisons. Tetrad V, comprising the eastern half of Yarner Wood and the adjoining woods in the Bovey valley, had 22 singing males, and Tetrad Q, comprising the western half of Yarner Wood, had 14 singing males. In the Dart valley, Tetrad E, comprising Holne Woods, New Bridge and the western half of Holne Chase, had 21, and Tetrad J, comprising the eastern half of Holne Chase, Ausewell Wood and Holne Bridge, had 18 singing males. Tetrad D, comprising the area of the tributory West Webburn River as far north as Ponsworthy, had 11 singing males.

The only other 10 kilometer square on the moor that approached the density of SX77 tetrads, was in SX59 where, in the Okement woodlands around Okehampton, 37 singing males were found in six tetrads, giving an average of 6.17 per tetrad. The only other square surveyed in the Dartmoor area with a high count was SX69, including Sticklepath and Throwleigh, where 26 males were found in nine tetrads, although a portion of these could have been outside the Dartmoor boundaries. The density here though was only 2.88 per tetrad, just lower than 3.19, the county average during the survey.

The valley of the Teign has much suitable habitat, and has held a good population for many years, although not covered in the random squares of the 1984/85 survey. The first mention of birds here in the *Devon Bird Reports*, was in 1937, when they were reported at Steps Bridge, although they were no doubt present many years before this. Dunsford Wood had up to 11 singing males in the 1990s, but the total had reduced to four by 2002. The area around Fingle Bridge and Drewsteignton had a good population also in the 1990s with 15 singing males in 1994 and 17 in 1997, but here as well numbers were reduced, with five recorded in 2001, and only one in 2002. The Castle Drogo estate had six males in song in 2001, and even woods around Fernworthy Reservoir have had a small population, first noted in 1961, and peaking at six in 1996-98 and seven in 1999.

Yarner Wood has always been much favoured by Wood Warblers. The first mention was in 1958, and by 1979 26 singing males were counted. The population stayed at around 30 to 35 from the early 1980s to 1999,confirming the figures of the BTO 1984/85 survey, after which, again there was a drop in numbers. A little further up the Bovey valley 12 singing males were recorded in Houndtor Wood in 1989, and 10 in Lustleigh Cleave in 1987. Even small areas of suitable wood can hold breeding birds, and the three areas of relict Dartmoor oak wood have all had breeding birds. Wistman's Wood has not had records in recent years, but Black-a-tor Copse, where the first bird was recorded in 1956, still has a breeding pair or two, and up to four have been noted at Piles Copse.

One interesting habit of Wood Warblers at certain Dartmoor locations is their adaption to conifer plantations. In the 1984/5 BTO survey birds were only found in 11 tetrads of conifer woodland in the county, amounting to only 2% of all records. However, the main area where this phenomenon occurs, the plantations of Soussons and Bellever, were not covered in the survey. In six of the 12 years from 1956 to 1967 up to five singing males were recorded in the study area around Postbridge, mainly in beech copses and woods in the West Webburn Valley, Bellever and Archerton. In 1961 a pair bred in Bellever plantation under two beech trees, amidst the conifers (Dare and Hamilton 1968). By the early 1990s birds had increased, and were found at about 20 sites each year, including 13 singing males in the mature open spruce forest, where breeding was strongly suspected (Dare 1996). This acceptance of conifers was also noted at Burrator from the mid-1970s, and has occasionally been recorded from plantations on Bodmin Moor in Cornwall, a county where much suitable deciduous woodland is without the species. In 1996 a male took up territory at Burrator in a quite dense, rather low canopied larch stand, with bluebell ground cover.

The Breeding Bird Survey index has shown a decline of 43% nationally for Wood Warbler, in the five years 1994/98 (Mead 2000). This appears to be reflected in certain Dartmoor sites from about 2000, but perhaps this is more a natural fluctuation, as records for the Dart valley, available from data collected for the Devon Atlas Project in 2002, would appear to indicate. Here the area, mainly on the south bank of the Dart between Dartmeet and Holne Bridge, produced a very healthy 82 singing males in oakwoods. An area including Holne Wood, Holne Chase and North Park Wood alone held 48 males, well up on the 39 males in an almost identical area in 1984/5. Future survey work in the county, including Dartmoor, will no doubt clarify the position.

Wood Warblers arrive back on breeding territories from the middle of April, with the main arrival in late April and early May. Exceptionally early arrivals have been at Belstone on 2 April 1990 and Yarner Wood on 7 April 1999. After the period of male song ends, usually in June, they can be very difficult to locate. After fledging, adults and young seem to fade away, with very few records after the end of July. There have been only about seven records of birds in August in the last 50 years or so, most of them in the first few days of the month. These records are, almost without exception, from breeding areas, and are thus more likely to be late departing breeders than migrants. The latest record, and the only one in September, was a bird at Yarner Wood on 5 September 2000. This species is very enigmatic as far as its status as a migrant in Britain is concerned. It is very rarely seen on the coastal migration hot-spots in spring or autumn, and the mass of the population appear to come straight in onto breeding territories, and in late summer depart again, as unobtrusively as they arrived. It is therefore interesting to have at least one record of a migrant away from any known breeding area. This concerned a bird trapped at Sampford Spiney on 27 July 1980.

CHIFFCHAFF *P. collybita*
Migrant breeder and passage migrant. A few winter.
This is a species that needs to have tall deciduous trees in its breeding territories. It does not take to dense conifer plantation, and does not favour young woods and scrub, so readily accepted by the Willow Warbler. Due to these habitat requirements, the population on Dartmoor is a little restricted, and is by no means as widespread as the Willow Warbler.

Around the Postbridge area in the 1950s and 1960s Chiffchaffs were very scarce, and the total breeding population was a maximum of five pairs, all of which were to be found in tall spruce on the edge of Bellever plantation (Dare and Hamilton 1968). By the early 1990s, although still scarce, they had increased to about 15 pairs, with five in deciduous copses, and 10 in the conifer forests (Dare 1996). There has been little other systematic coverage of sites over the years, but the Harford Common Bird Census site has had breeding birds most years from 1987 to 2000. Two or three territories have been normal here, but there have twice been years, 1994 and 1999, when birds have been present without establishing territories. Conversely there have been especially good years. Four territories were established in 1997, and six in 1998. On the eastern edge of the Dartmoor area, Dunsford Wood Nature Reserve had a stable population of four or five territories each year throughout the 1990s. Birds were very numerous in 2002 when nine territories were located.

Recently the fieldwork for the Devon Atlas Project in 2002 has shown some interesting results. Certain tetrads have shown good results, with 11 singing males at Fernworthy, ten at Dendles Waste, and eight on the Hunters Path in the Teign valley, and in the Bonehill area. The opposite has also been shown to be true in some areas. The woods of the Dart valley do not appear to be well suited to this species. Only one was found in the oakwood area below Bench Tor, and only a single singing male was located in the whole of Holne Chase.

Chiffchaffs are amongst the first migrant species to arrive in the spring. Arrival dates vary from year to year depending on weather conditions encountered on passage. An early year, when conditions are right, can see birds arriving in early March, as in 1997, when the first birds were encountered between 8 and 14 March, or 2000 when several were at Piles Copse and New Waste on 5 March. A more normal year would see the first arrivals in the third week of March, and a late year may not produce the first birds until late March. The main arrival is in April, and at this time migrants will come off passage for a time, join the local breeding birds and swell numbers. Occasionally at this time "grey" birds, probably of one of the Scandinavian populations, may be encountered. One was at Okehampton on 17 March 1989, and another was seen at Throwleigh two days later on 19 March. High counts in early April have included 38 at Kennick Reservoir on 4 April 1993, and 31 there on 9 April 1995. After breeding, birds tend to disperse from late July, and migrate south from late August, with a peak in the second half of September (Wernham *et al* 2002). On Dartmoor birds in small groups, sometimes foraging with other species, are seen regularly throughout September, but by early October numbers have declined and birds that are still about are usually seen singly. The latest October bird was a single at Welltown, Walkhampton on 29 October 1996. Birds will often sing in the autumn, and this has been noted as late as early October.

By early November the picture gets a little complicated. The few records at this time could refer to very late migrants, or they could refer to the arrival of wintering birds.

Wintering Chiffchaffs are not a new phenomenon. D'Urban and Mathew quote several references to birds in Devon, and indeed wondered if the bird they heard at Chagford on 9 March 1872 had wintered in the area. Dartmoor records are few however, as most winter birds stay in lowland habitat, where the temperatures are moderate, and insect prey more easily available. Most records have come from the Buckfastleigh area in recent years. There the local sewage works is a big attraction for birds. Single birds have been noted around Buckfastleigh in January 2001 and December 2002, but eight were at the sewage works on 16 January 2000. At other localities a bird was seen at Whitchurch on 22 February 1967, another at Sampford Spiney on 26 December 1980, and another that wintered there in 1982. A bird was at Burrator on 23 January 1983, and the single at Cadover Bridge on 20 November 1999 was more likely to have been a wintering bird than a very late migrant.

WILLOW WARBLER *P. trochilus*
Migrant breeder and passage migrant.
The Willow Warbler is a common, even abundant, summer visitor to much scrub and woodland habitat around the moor edge. Young conifer plantations are particularly attractive to them, although numbers decline when the canopy closes over. In

Soussons and Bellever plantations in the 1950s and 1960s, birds took full advantage of the young trees, and it was estimated that 250-500 pairs were present (Dare and Hamilton 1968). This number had greatly reduced by the 1990s mainly due to the maturing of the trees (Dare 1996). There was a definite reduction in the population nationally in the early 1990s, and this could have contributed to the low numbers in the plantations. It certainly contributed to the total absence of birds during a survey of the relict oakwoods of Wistmans Wood, Black-a-tor Copse, and Piles Copse in 1993 (Smaldon 1994). Six territories had been recorded in Piles Copse only a few years previously in 1989. This marked decline, which appeared only to affect populations in southern Britain, could well have been caused by poor adult survival due to problems on migration or in the African winter habitat (Mead 2000). A general recovery took place in the late 1990s, but at Dunsford Wood, where an average of 10.2 territories had been normal up until 1995, numbers gradually fell until only five were found in 2002. This may have been due to local circumstances, as the population at Harford Common Bird Census site, although having some fluctuation, was stable throughout, averaging 11 territories from 1987 to 2000, with a high of 17 in 1990. Amongst the records collected for the Devon Atlas Project in 2002 were 30 singing males in the Bonehill area tetrad, 29 at Fernworthy, 20 on the Hunters Path and 13 at Hexworthy.

The first arrivals in spring are usually in the first two weeks of April. Occasionally they can be earlier, with several records in the last week of March, and the earliest being three at Cadover Bridge on 10 March 2000. The main arrival is during the latter half of April and early May. During this time concentrations of birds will arrive in an area, only to move on very quickly. This was noted on Roborough Down on 5 May 1936, and again there on 15 April 2003. As with many species, they do not tend to be reported very much after the breeding season, and the post-breeding dispersal gets very little coverage. Most young have fledged by the middle of July, and migration proper takes place during August and September. During the autumn migration birds can turn up in some unusual places. One was searching a single Rowan in the Plym valley near Plym Steps on 2 August 1998. Even stranger was a bird, also searching a Rowan, surrounded by open moorland near Eastern White Barrow on 10 September 2001. Most migrating birds have departed by mid-September, but records in the latter half of the month are not too uncommon. October records are almost unknown, with three birds at Golden Dagger Mine on 6 October 2002 being the only one. Amazingly there has been a November record, with a bird giving sub-song at Okehampton on 2 November 1989.

GOLDCREST *Regulus regulus*
Resident breeder and possible passage migrant.
This tiny mite is always grossly under-recorded. It inhabits the moorland plantations, where it can be quite common, but because of its thin song and habit of staying high in the canopy, very few records bear much relation to its true status, and numbers

present. Away from the high densities in mature conifer plantations, Goldcrests also inhabit mixed shelterbelts and copses, and even gardens where a mature conifer or two are present. Since the planting of conifers on moorland hillsides they have increased dramatically. However, of all woodland species, they are probably the hardest hit by severe winter weather.

Up until the 1962/63 winter it was estimated that 250/500 pairs bred in the plantations around Postbridge, and in the scattered copses and shelterbelts in the area. After that severe winter the number had reduced by 95%. The recovery, however, was swift, with numbers more or less back to the previous level within three years (Dare and Hamilton 1968). A similar depletion and recovery, was noted in other areas of the moor, as in Devon generally. Any period of cold weather, even a relatively short one, where glazed ice persists on trees and foliage is catastrophic for this species (Sitters 1988).

Four or five pairs on territory at the Harford Common Bird Census site 1987 to 1990, were reduced to none in 1991, probably as a result of the arctic conditions of February that year. Recovery was not too speedy, and it was 1995 before four territories were again occupied. Another drop in numbers occurred in 1996, when birds were present but no territorial behavior was noted. There was a short period of severe cold in December 1995 that could have had an effect on the birds at this relatively high site. It again was another four years, in 2000, before four pairs were back on territories. Dunsford Wood Nature Reserve, in the valley of Teign, much lower in altitude and more sheltered than Harford saw no such reductions in population. The average here between 1991 and 1995 was 3.5 pairs/territories. It rose to five in 1996, six in 1999, and eight in 2000. Very useful survey work was carried out for the Devon Atlas Project in 2002. In areas that were away from the prime conifer plantation habitat, nine singing males/pairs were located in a part of Holne Chase (in SX7270), eight at Huccaby, six in the West Dart valley north of Hexworthy, four in the Bonehill area, North Park Wood, and Throwleigh Common.

After the breeding season family parties will join together, at times with other species, and roam the area in search of food. Small numbers are seen in many places but larger numbers have been noted at Fernworthy where a post breeding flock of 40 was seen on 23 August 1988, and 32 on 20 August 1997. In September and October there is the possibility that some birds seen could be migrants, rather than dispersing local birds. This is certainly the case on the coast, but inland it is rather more difficult to prove. One in a single Rowan near Eastern White Barrow on 10 September 2001, was in company with a Willow Warbler, and could well have been a migrant. A noticable passage of birds was seen at Meldon Reservoir on 3 October 1988, and a similar influx was reported at Fernworthy on 23 September 1997.

As autumn drifts into winter, birds are forced to feed outside their normal habitat on occasions. One of the main types of habitat exploited on Dartmoor at this time of year is that of gorse scrub and bracken. At Mountsland Common on 30 December

1989, 12 were disturbed from dead bracken at dusk, suggesting that birds will not only feed, but also roost in this unusual environment. At least 20 were feeding in bracken at Aish Tor on 29 December 1999, and at least 50 were there on 15 November 2000. Birds have also been noted in gorse and bracken at this time of year at Roborough Down and Huccaby Cleave. Exploitation of a different habitat was recorded at Buckfastleigh on 27 January 2000, when about 50 were seen around the sewage works.

FIRECREST *R. ignicapillus*
Winter visitor and passage migrant. May have bred.

The first record from the Dartmoor area involved two at Owley, Wrangaton on 20 December 1967. This was during a time of expansion for the species and it was expected that birds would probably breed in the county during the 1970s. With its abundance of conifers and also oak woodland with holly understorey, Dartmoor seemed well placed to accommodate a pair or two. Unfortunately, this has not as yet happened.

There have been four instances of possible or even probable breeding. The first was a singing male found at Hennock Reservoirs on 30 April 1982. To date the most likely breeding record, was a pair seen mating on 7 June 1990 at Harford Common Bird Census site. The birds were not seen again that year, although birds were present in the 1996 breeding season. In 1994 a male was trapped at Bridford on the interesting date of 1 May.

Excluding these territorial birds, there have been 47 records of birds on Dartmoor up until the end of 2002. Five have been within the March spring migration period, 17 have occurred in the normal autumn migration period of late September to early November, but rather surprisingly most, 25 in total, have been found in the winter months of December to February. This would tend to suggest that the birds seen at Dartmoor locations have dispersed there after arriving at more lowland or coastal sites the previous autumns. However, observation in certain areas of the Dart woodlands in recent autumns has turned up birds fairly regularly, including the first record of three together, so it is possible that birds could have been missed in the past.

There is no real pattern as regards localities for Firecrests. They can turn up just about anywhere, provided sufficient cover is present. The majority of records, though, have been at locations around the periphery of the moor rather than the interior. Birds are particularly linked with river valley areas, and the woods that usually line these valleys. Over 30 of the birds recorded have been around the southern edge of the moor, with the valley of the Dart having about 15 of these records. Another group of records come from the valleys of the Meavy and the Plym, including Burrator. On the north side of the moor, only Fernworthy has had several records. The top spots have been Buckfastleigh with eight birds, and Fernworthy with six.

SPOTTED FLYCATCHER *Muscicapa striata*
Migrant breeder and passage migrant.

The numbers of this summer visitor have decreased drastically in Britain in the last 30 years, possibly by as much as 75%. What was once quite a common bird has become difficult to locate at times. The last few years have shown some signs of an increase, so perhaps the worst is over.

On Dartmoor there have been insufficient records to assess the position fully, but anecdotal comments reflect the national decline. In the Postbridge area it was widely distributed in the 1950s and 1960s, breeding around most habitations, as well as in some copses. A total of 15 to 25 pairs were present (Dare and Hamilton 1968). By the early 1990s the figure had reduced to about 10 pairs. At the Harford Common Bird Census site, two pairs were regular each year from 1987 to 1993, but only one pair was present from 1994 to 2000. Three or four pairs were normal at Dunsford Woods from 1989 to 1992, but by 2002 this total had reduced to a single pair. Birds have been noted breeding in the three relict oak woods in recent years, with Wistman's Wood, Black-a-tor Copse and Piles Copse each having a pair.

Although birds shun open habitat for obvious reasons, they will on occasion accept the bare minimum of tree cover for flycatching, and ruined buildings for a nest site. The pair that have been noted at Teignhead Farm in 1998, 2000 and 2002, prove this point. Field work for the Devon Atlas Project in 2002 showed the species to be well distributed, but in no great numbers. About 19 pairs/territories were found in the Dart and West Dart valleys between Huccaby and Hembury Woods, including four at Holne Chase and three at Huccaby, where the same number had been noted since 2000.

Spotted Flycatchers are late migrants, with the bulk of the population only arriving in mid-to late May. Normally the first birds to arrive reach breeding territories during the first two weeks of May, but in the last nine years there has been a trend towards earlier first arrivals, with nine in April. Prior to 1995 the only April arrival noted on Dartmoor was a bird found dead at Ashburton on 29 April 1932. From 1995 the nine recorded have been mostly in the last 10 days of the month, but one was at Meldon Reservoir on 19 April 2000, and one incredibly early individual was at Harford on 1 April 1999. After the breeding season, family parties disperse around the natal area prior to migration. At times families join together to form roving bands. Ten to 15 were working the hedges as a group near Clearbrook Cross on 4 August 2001, and at least 16 were at Moorhaven Hospital, Bittaford on 23 August 1987. The largest of these groups, containing about 20 birds, was seen at Two Bridges on 5 September 1989. The latest any of these groups have been noted is 8 September, and the birds seen later than that, probably migrants from elsewhere rather than local birds, are usually seen singly. The latest migrants have been birds at Moorhaven on 22 September 1986, and at Lettaford on the same date in 2000.

RED-BREASTED FLYCATCHER *Ficedula parva*
Very rare vagrant.
The only bird on Dartmoor was a male, still in summer plumage, at Devon Friendship Mine, Mary Tavy on 27 September 1958. It was perched on a stay wire, and was well described by the finder, being seen at only six yards range. Of interest was another or perhaps the same male, seen on 11 September the same year just outside the Dartmoor boundary, at Fillham, Ivybridge.

PIED FLYCATCHER *F. hypoleuca*
Migrant breeder and passage migrant.
The records relating to the fortunes of the Pied Flycatcher on Dartmoor, and the story of the various nest-box schemes that encouraged its spread, would almost fill a book by itself. So it will be necessary to give only a rather brief outline here.

Late in the nineteenth century D'Urban and Mathew considered the Pied Flycatcher a casual summer visitor to Devon. It was rumoured to have bred in North Devon, but in many of their forays onto Dartmoor to look for them, the authors had found none. Rather later, in 1945 a pair was seen at Hexworthy on 10 and 12 May. They were thought at that time to be migrants, but perhaps a breeding attempt was made close by. A pair did breed in a hole in a barn wall at Scorhill near Chagford from 1949 to 1951.

In 1952 Yarner Wood was one of the first six National Nature Reserves to be designated in the country. Dr. Bruce Campbell surveyed the wood in 1954 and recommended the establishment of a nest-box experiment. On re-surveying the wood in 1955, it was found that one of the new nest-boxes held a pair of Pied Flycatchers. They were the most southerly nesting pair in Britain (Page 1992). The colonisation of Dartmoor's oakwoods had begun.

After the initial breeding at Yarner, numbers built up gradually over the next few years. Six nest-boxes were occupied by 1959, and although there was some fluctuation in yearly numbers, nine boxes were being used in 1966. Then came a sudden increase, with 18 boxes used in 1967, 26 in 1968 and 20 in 1969. There was then natural fluctuation over the next 10 years, with the number of occupied boxes averaging about 14 per year. This again increased suddenly to 24 in 1980, and then rose steadily to the present day, with numbers varying between 30 and 70 annually. The average was 49. The worst years were 1981 and 1983 with 30, and the best year was 1990 with 70 boxes occupied. After the establishment of the Yarner colony, further boxes were put up in the Bovey Valley woodlands. Here too, birds were quick to take advantage of the nest sites provided. A pair used a box in Knowle Woods, Lustleigh in 1969, with pairs at Becky Falls and Houndtor Wood from 1973 and 1977 respectively. Eighteen boxes were used in Bovey valley in 1984, with 7 boxes used in Higher Knowle Woods in 1987, and 4 in the new scheme at Neadon Cleave in 1987. The nest-boxes at this last site have been monitored annually since 1987, averaging about seven occupied per year, with a maximum number of ten in 2002.

Once the Yarner Wood experiment had proved a success, other woodland sites followed suite with nest-boxes. One of the first was at Wrangaton where two pairs bred in boxes in 1967, and one or two pairs were present intermittently up until 1991. Three pairs bred in 1980. In the valley of the Yealm at Dendles birds were first noted in 1968. Although not surveyed in every year, the site had five boxes being used by 1970, and counts there went into double figures in the mid-1980s, reaching a peak of 21 in 1991. The woods in the valley of the Taw around Belstone, Skaigh and Sticklepath were first occupied in 1970 when a pair bred in a back garden of a house in Belstone (Vaughan 1979). Although numbers in this area were never very high, perhaps due to limited habitat, a maximum of eight pairs were present in 1986.

The establishment of birds at Belstone was the spur for the start of the only nest-box scheme on Dartmoor to rival Yarner in size. In 1971 a pilot scheme was commenced to erect boxes in wooded valleys around Okehampton. After three years of false starts and site disturbance, a quieter area was found, and birds first bred in 1974. That year two pairs bred in boxes, and a third at a natural site. As elsewhere, once established, numbers quickly increased with 11 boxes being used by 1977, and a huge increase to 22, with others at natural sites, in 1978. Numbers since then increased, with the addition of many extra nest-boxes, and a maximum of 68 boxes were used in 1989, thus peaking a year earlier than Yarner. Since then numbers have decreased somewhat, although 2002 was a good year with 59 boxes occupied (Vaughan 1979 and 1994). Boxes were erected at Burrator in the mid-1980s, after the first attempted breeding here in a natural tree site in 1978. Success was limited, but increased greatly when the boxes were relocated to a new site near the dam in the early 1990s. The peak year proved to be 1996 with 16 boxes used. Since the late 1990s a small number of pairs, maximum three, have bred in additional boxes in the Deancombe valley. The Teign valley has provided several nest-box schemes to assist the spread of Pied Flycatchers. Dunsford Woods has had an increasing population since the provision of boxes in the early 1980s, with the maximum of 22 pairs in 1989, and Bridford Woods had 18 pairs more recently in 1995. On the west side of the moor, a scheme on the River Walkham at Grenofen from 1985, peaked with 14 boxes in use in 1994. Over the years there have been several other breeding schemes, mostly small, that have augmented the totals.

As can be noted from the above, the years from 1988 to the mid-1990s were very good as regards the numbers of breeding birds using nest-boxes around the moor.

However, occupied boxes do not always equate to good breeding success, and the cold, wet, late springs recently have certainly not helped matters. In addition to this, the problems of resident Blue and Great Tits claiming boxes before the migrant flycatchers arrive, lack of food when a bad year for defoliating caterpillars occurs, weasel predation, and even the occupation of boxes by dormice must be taken into consideration, and it can be seen that the picture is not all rosy. Although the general feeling must be one of optimism, these delightful birds are very heavily reliant on man-made nest sites, and so their continued welfare and expansion may rest on the continuance of the nest-box schemes begun so successfully at Yarner Wood 50 years ago. One piece of information that came to light recently after extensive survey work for the Devon Atlas Project in the Dart valley was the location of 42 pairs of singing male Pied Flycatchers between Dartmeet and Holne Bridge. All of these were in natural sites without nest-boxes, and in view of this perhaps birds are becoming established in woodland normally away from the public eye, and our breeding population may be substantially larger than the total of occupied nest-boxes would suggest. Time, and more survey work, will no doubt provide the answer.

There is unfortunately little space here to say much about the numerous records of birds ringed, re-trapped or otherwise recovered in our area. Suffice it to say that these records are very numerous, and annually the total increases. Many birds are site faithful, but there are numerous instances of individuals moving between different sites on the moor. Birds have also come into the area to breed after being ringed in Cornwall, Somerset or Wales in previous years, and Dartmoor-ringed birds that have been re-trapped in these other regions. An example of this was a male that bred at Wrangaton in 1972, having bred the previous year at Lanhydrock, Cornwall. This bird was especially interesting, as having left Dartmoor to journey south in the autumn, it was retrapped in Portugal on the 4 October 1972. There has been one case of a pullus ringed at a moorland site being found dead in France in August on its first migration south, five instances of birds being ringed as pulli found dead in Morocco, on their spring migration north in subsequent years, and one record of a female ringed as an adult at Okehampton on 6 June 1989 and found dead in its winter quarters in Guinea, West Africa on 16 January 1991.

Pied Flycatchers arrive in spring from mid-April through to early May. Arrival times at breeding sites depend greatly on the weather conditions encountered on passage. In a year with favourable conditions, the first birds can arrive during the first week of April, the earliest ever being a bird at Yarner Wood on 28 March 2002. When conditions have been against them the first arrivals are much delayed and at times do not appear until near the end of the month, such as in 1978 when the early birds were only at Okehampton on 23 April, Bovey Valley on 26 April, and Yarner on 27 April. After breeding, as with so many species, little is known of their movements. Many disperse around the area, at times with tit flocks prior to migration. The last birds at Yarner usually depart in the first week of August, and the few records of birds in late August and September no doubt refer to migrants passing through the area. There have been three early September records, with the latest being at Belstone Cleave on 9 September 1997, but the latest bird was one at Meldon Woods on 25 September 2002, incidentally the same year that saw Dartmoor's earliest spring record.

LONG-TAILED TIT *Aegithalos caudatus*
Resident breeder.

The Long-tailed Tit is a species requiring dense scrub, bushy hedgerows and woodland edge habitat in the breeding season. If it is in open woodland there has to be a good shrub layer to make it acceptable. It tends to be a bird of lower elevations than other tits, and as such its presence on Dartmoor is largely confined to river valleys (Sitters 1988). It does not travel far from the breeding area during the year, so can be badly affected by any severe weather in the winter. In the Postbridge area, where they have always been quite scarce, about five pairs breed in a good year, but this figure is reduced to zero after severe winters (Dare 1996).

Numbers tend to pick up in a few years following a bad winter, and there appears to have been no change in status in recent times. Recording of numbers has not taken place at many sites, but at Harford Common Bird Census site a good year with three territories in 1990, was followed by a single pair breeding each year up to 2000, with the exception of 1996 when the birds did not appear. The Dunsford Wood Nature Reserve was at a low of only one occupied territory in 1992, but this had increased to six by 1994, and seven, the highest since 1980, in 1996. The latest report was of up to four pairs in 2002. At Yarner Wood, where the species breeds regularly, four pairs were recorded in 1992.

After the breeding season, family parties, although perhaps not travelling any great distance, roam around the local area. Families will link up with others in their search for food, and at times these parties can contain a substantial number of birds. Ten to 20 in a feeding party is not unusual, and parties of 30 have been noted several times. Forty five in the West Webburn valley on the early date of 19 June 1957 were exceptional, especially so as the species is far from numerous in this area. The maximum counts have been 50 in several smaller groups at Burrator on 10 November 2002, and 52 at Okehampton on 22 January 1989.

These roving flocks can be encountered any time from the end of June through to the following March, although as an early breeding species most birds have re-established their breeding territories by February or early March.

At times these flocks, their courage no doubt boosted by their numbers, will move right out of their normal habitat and feed in a much more open area. Examples of this include the 12 searching hawthorn bushes below Black Tor on the Meavy on 7 January 1990, the 10 in willows high up the O Brook near Hooten Wheals on 20 October 2001, and the 12 moving between hawthorns in open country on Buckland Common on 28 October 2002.

MARSH TIT *Parus palustris*
Resident breeder.
The name of this bird is misleading, as it is not confined to wetland sites, but is a bird of mainly mature broadleaf woodland with a scrub understorey. The Dartmoor distribution of this very sedentary species reflects its preferred habitat. Although widely distributed in suitable woodland, its numbers have never appeared great, and indeed the 2000 Dartmoor Moorland Breeding Bird Survey found them very scarce around the periphery of the moor (Geary 2000).

Yarner Wood has had more records than any other location, and since 1958 has had pairs using nest-boxes, something Marsh Tits are not generally known for. Even here numbers have never increased, and the highest number recorded would appear to be the six pairs noted in 1954 (Page 1992). Since then, although the published figures may not be the actual totals present, the highest figure recorded has been three pairs in 1999. The use of nest-boxes has also been noted in Bovey Valley Woods and Neadon Cleave, but at Okehampton the pair that bred in a nest-box in 1989 was the first for 18 years. At Burrator, where pairs have been present in at least three areas since the mid 1990s, a pair was noted investigating a nest-box in Burrator Wood in 1999, but whether it was actually used is not known. Dunsford Wood Nature Reserve had two or three pairs between 1994 and 1996, but Harford Common Bird Census site had only one record of a pair present in 1987. The latest fieldwork, for the Devon Atlas Project in 2002 showed birds to be present, if somewhat thin on the ground in the oakwoods of the Dart valley. Three pairs were found in a part of Holne Chase, two pairs, with probably more present, in Holne Woods, two probable pairs at Gallant le Bower, and one or two pairs at Huccaby.

Marsh Tits are not known to frequent gardens and other areas close to human habitation, as much as the commoner *parus* species. However, as long ago as 1938 they were recorded as being the commonest tit in winter at H.G. Hurrell's home at Wrangaton, although the nearest known nest was about a mile away. Recently they have been noted regularly visiting a garden at Wootton, Holne in winter, and in 2002 the first birds started visiting in the autumn on 28 September. Garden visiting, mostly attracted by feeders, has also been reported from Buckfastleigh, Lydford, Lustleigh and Okehampton, where 1986 was the first time in 20 years that none had been seen in a garden.

As would be expected from a species that moves very little from its breeding territory throughout the year, large gatherings are not usual. In autumn and winter groups of more than six together are rarely reported, and the largest was only 10 at Burrator on 22 February 1984.

WILLOW TIT *P. montanus*
Scarce resident breeder.
Nationally this is a declining species, but although undoubtedly scarce on Dartmoor, birds could well go unrecorded. Many reports have come from sites that had previously recorded birds 10 or even 20 years before. It must be assumed that these sites were not looked at well in the intervening years. Being the only *parus* species that regularly excavates its own nest hole, it is heavily reliant on a good supply of soft rotting timber, tree stumps, etc., for nest sites. It therefore favours damp mixed woodland mainly of birch, willow and alder, and on Dartmoor, conifer plantation areas that have plentiful dead wood.

Thirty four sites have had past records, but three hold nearly half the of the total:

Fernworthy Reservoir : Birds were first noted here in 1971, when at least two pairs were present. Although no doubt still in the area, birds were not reported again until 1978. Then commenced almost annual sightings throughout the 1980s and up until 1993. Single pairs were confirmed breeding in 1987 and 1991, and two pairs bred in 1988. They were recorded again in 1999, and every year since.

Hennock Reservoirs : First recorded in 1979, and then in 1980 and 1984. Then eight years passed without records until 1992, when yearly records were noted until the present. Although 1995 was the only year when breeding was proved, it must have taken place most years from the early 1990s. Two pairs were seen at Kennick Reservoir in 1993, and it would appear that birds are widely, if sparsely, distributed in the area, with reports from Hennock, Tottiford Reservoir and Laployd Plantation over the years.

Burrator : As at Hennock, birds were first noted here in 1979, then noted annually around the reservoir until 1988. There has been no confirmed breeding record, but birds were seen entering a probable nest hole in 1986. There was a breeding season record in 1991, with birds seen also in 1994 and 1997, but no confirmed sightings since.

Confirmed breeding has also occurred at Belstone in 1964, near Chagford in 1975, Meldon also in 1975, Okehampton in 1980, Throwleigh in 1990, and probably in Yarner Wood in 1999. As the species is largely sedentary, it is probable that most records refer to potential breeders irrespective of the time of year. However birds can appear in areas where they have only rarely, if ever, been reported before. One at Wrangaton on 1 November 1964 was the first there for 25 years, and the bird trapped in Sampford Spiney on 5 July 1980 was the first in the area, although another was found in 1990.

A survey of Rhôs pasture for the DNPA in 2000, found five territories in four systems, two at Widecombe, and single territories at the Erme Valley, Easton, and Tor Valley. There is clearly great potential for unrecorded pairs in Rhôs pasture areas around the moor edge.

COAL TIT *P. ater*
Resident breeder.

Coal Tits must always have been regular and widespread in the deciduous and mixed woodland around the moor edge, but with the planting of the coniferous plantations in the early twentieth century they really came into their own. Conifers are prime habitat for Coal Tits, and with the maturing of trees the population has grown immensely.

At Postbridge in the 1950s and 1960s, between 150 and 250 pairs were estimated to have bred, almost all of them in the plantations at Bellever and Soussons. The population was severely hit by the bad winter of 1962/63 and the mortality was thought to be about 75%. Numbers increased again very quickly and had reverted to former levels by 1965 (Dare and Hamilton 1968). There has been no recent assessment of population there, but it is likely to be higher than the totals of the 1950s/1960s.

Although badly hit in 1962/63, it is questionable if the population has been affected by any of the less severe winters since. They are not thought to suffer from the effects of prolonged cold as much as some other small species, because of their habit of storing food and their ability to feed on the underside of conifer fronds, away from any ice and snow. The overall numbers at present appear high, and the population seems to be stable or perhaps still increasing.

At the Harford Common Bird Census site, occupied territories have ranged from three to 11 during the years 1987 to 2000, with the average of seven. The best years, with 11 territories, were 1991, 1992, and 2000. Dunsford Woods Nature Reserve had a high of 16 territories, also in 1992, but numbers fell to five in 1993 and 1994, and had only risen to seven by 1999. Recent estimates of population have included at least 11 singing males at Huccaby House in 2001, four pairs/singing males in Huccaby Cleave in 2002, about five at Stennants, Hexworthy the same year, and four pairs around the edge of Yennadon Down also in 2002. Coal Tits will use nest-boxes on occasion, but on Dartmoor the habit appears rare. The pair that fledged seven young from a box at Okehampton in 1998, was the first to do so since the provision of the boxes in 1972. At Yarner Woods a pair used a nest-box in 1968, and 1988 and two pairs in 2001. Burrator Wood had a pair in a box in 1999.

Although birds are common in plantation areas during the breeding season, they usually keep to the canopy, and are heard more than seen. Once the young have fledged they are much more in evidence, and from July family parties will join together to locate suitable food sources, sometimes with other *Parus* species. These parties are then a feature until the next spring, feeding in conifers, on beech mast under trees, occasionally working through patches of gorse. When weather is inclement they have

Coal tit

no hesitation in descending on gardens and bird feeders. About a third of the larger parties recorded have been seen in December. These have included about 40 together at Fernworthy on 3 December 1997, and the highest count of 58 again at Fernworthy on 9 December 2002.

BLUE TIT *P. caeruleus*
Resident breeder.
The 1937 *Report of the Devon Bird Watching and Preservation Society* carries a statement by Malcolm Spooner, who had surveyed the plantations on the moor, that Blue Tits were very scarce or absent at higher levels even in admirably suited habitat. Twenty or more years later, in the Postbridge area they were widely scattered in sheltered deciduous woodland and around habitation but the total population only amounted to 25 to 50 pairs (Dare and Hamilton 1968). By the early 1990s only 10 to 15 pairs could be located (Dare 1996). The reason for the decline was not known. It would appear that central areas of Dartmoor, even if likely habitat exists, is not well suited to this species.

The same cannot be said of the deciduous woods and valleys of the moor edge. Here, no doubt greatly assisted by the several nest-box schemes, the species thrives. The Blue Tit is very much taken for granted by many people, and thus casual and even more professional recording is very scant. One exception to this is the Common Bird Census site at Harford, where records have been kept from 1987 to 2000. The population here has remained stable with an average of 21 territories in a 76.7 hectare plot. The maximum in that period was 29 territories in 1991, and the minimum 15 in 1997. For most data on the Blue Tit we must look briefly at the records of nest-box schemes. The largest two give an indication of just how important they are for the species:

Okehampton Woodlands: The nest-box scheme was started in the early 1970s primarily to attract Pied Flycatchers, but Blue Tits, as everywhere, were attracted to the boxes. By 1980 25 boxes were being used, and by the end of the 1980s this figure had risen to 62. Numbers rose more gradually during the 1990s, and a peak of 77 was reached in 1998. The new millenium saw 76 used and a record 81 were occupied in 2002. From the mid-1990s the success rate was greatly affected by the cold, wet spring weather. The rate in 1993 was the worst for 20 years, and 1996 was the worst ever, as it was nationally. On an optimistic note the success rates for 2000 to 2002 were very much higher, with 94% fledging in the latter year.

Yarner Wood: Here the rise in numbers using boxes has been a little more erratic than at Okehampton. Figures on record for the 1980s varied between 10 boxes in 1984, a poor year, to 40 the next year. In the 1990s the best year was 1996, with 58 pairs in boxes, and the lowest 1999 with only 34 pairs, but with a good breeding success rate and a large clutch size. In 2000 and 2002, 48 pairs were in boxes, and 52 were present in 2001. Success rate was not too good in 2000, but much better the next two years, with 88% fledging in 2002.

Other nest-box schemes have operated, and are mostly still operating at Bovey

Valley Woodlands, Neadon Cleave, Dunsford Wood, Burrator, Dendles and Grenofen.

Being largely sedentary, Blue Tits are not normally encountered together in large numbers, but in autumn some noteworthy gatherings have been 68 in the Glaze Brook Valley on 10 August 1990, 27 at Huccaby on 19 September 2001, and at least 40 at Burrator on 24 September 2001.

GREAT TIT *P. major*
Resident breeder.

A common bird in all suitable moor edge areas and wooded river valleys that reach into the moor like the Dart, but like the preceding species less common in certain central areas. The Postbridge district had a population that was thinly distributed throughout cultivated areas in the 1950s and 1960s, and numbered about 25 to 50 pairs (Dare and Hamilton 1968). By the early 1990s a considerable reduction had taken place and only about 10 pairs could be located (Dare 1996). Indeed in 1990 they were found to be particularly scarce, with only five territories found.

In the sheltered valley of the West Dart at Huccaby, Great Tits are one of the commonest birds (Hibbert 2000). As evidence of this at least 15 singing males were found in 9.3 hectares in 2001. As with Blue Tit most details of population come from the various nest-box schemes around the moor. Nowhere does the species occupy as many boxes as the Blue Tit, and at Yarner Wood and the other sites in the Bovey Valley numbers are low. Only in the Okehampton Woodlands do numbers approach that of Blue Tit. From the early days of the scheme there, Great Tits occupied between 30 and 44 boxes, with a high of 52 in 1985. The late 1980s saw a reduction to 20 to 32 boxes, but a high was reached again 1991 when 54 boxes were used. Numbers averaged about 36 for most of that decade, but rose to 49 in 2000 and an all time maximum of 63 in 2002. Yarner Wood, in comparison, only had an average of 8.2 boxes occupied yearly from 1988 to 2002, with Bovey Valley Woodlands and Neadon Cleave reaching only 3.1 average yearly during the same period. Of course, direct comparison is not really possible without the size of wood and the number of boxes present being brought into the equation, but it does appear that the drier eastern woodlands are not so productive for this species as the more western areas. In most areas the success rate of fledged young was higher in 2001 and 2002 than it had been for many years.

There have been few records of any numbers of birds together, but 15 feeding under beech mast at Grendon Cot, Soussons on 25 November 1997, was noteworthy.

NUTHATCH *Sitta europaea*
Resident breeder.

This highly sedentary species can be found on Dartmoor wherever deciduous woodland and copses occur. Pairs on territory stay in the area throughout the year, and only birds in their first year travel short distances. Occasionally a bird will move outside its normal habitat, as one watched searching through a hawthorn bush by Ingra Tor in May 1986. Their habit of food hoarding assists in their sedentary life style (Gibbons 1993). They have increased nationally over recent decades, and are possibly still doing so, although the picture on Dartmoor is not too clear.

There certainly has been an increase in the Postbridge area, where the three to five pairs of the 1950s and 1960s, had risen to eight to 10 pairs by the early 1990s (Dare 1996). Most other areas do not possess earlier records to make any comparison, but what recent records there are suggest a population that is at least stable. At Huccaby House, in the valley of the West Dart, the species is a common resident, with four territories close by (Hibbert 2001). They are regularly reported from at least four areas around Burrator reservoir, and during survey work for the Devon Atlas Project in 2002, seven territories were noted in a part of Holne Chase. At the Harford Common Bird Census site an average of 4.5 territories have been present between 1990 and 2000, with a maximum of eight in 1991.

As well as using natural tree holes, Nuthatches take readily to nest-boxes, and will reduce the entrance hole with mud if it should be too large. It is then again to the nest-box schemes that we must turn, for most information on the area's population. The boxes in Okehampton Woodlands have held a healthy population since their erection in the early 1970s. Ten boxes were being used in 1979, and throughout the 1980s the average was about the same. Fourteen boxes were occupied in 1990, with 20 the next year which to date is the highest number used at the site. This did not signal a prolonged increase however, and the average for the rest of the 90s was between 11 and 12. Recently 13 to 15 boxes have been used. The fledging success rate can vary considerably depending largely on the weather conditions, with cold and wet springs taking a heavy toll at times. 1986 and 1995 were particularly bad years with fledged young per box averaging at about 3.2, as against a figure of about six in good years.

The boxes at Yarner Wood averaged just over five occupancies a year from 1988 to 1994, with a high of nine in 1990, but then numbers crashed to only one in 1995, and have not averaged more than two per year since. Fledging success here though is usually very good. Bovey Valley Woodlands and Neadon Cleave have also never had more than two or three pairs using boxes annually.

Although normally using woodland nesting sites, be it natural or nest-box, pairs have also been recorded breeding in holes in walls, sometimes only a metre or so from the ground. A pair bred in the wall of a cowshed in the West Webburn valley in 1961 beside a much used doorway, in a barn wall about four feet from the ground at Wrangaton in 1982, and in a low wall at St. Michael's Church, Princetown in 2001.

TREECREEPER *Certhia familiaris*
Resident breeder.
This unobtrusive, mouse-like bird inhabits much deciduous and mixed woodland on Dartmoor, and even breeds in the conifer plantations. However, its nature is so quiet

that it gets noticed very little, and to those acquainted with its thin high-pitched song, it is probably heard more than seen.

As records are relatively few, it is not possible to give much information on its distribution and numbers, other than to say it is a widespread but low-density species, and is not scarce. During the 1950s and 1960s between one and five pairs bred in the Postbridge area, mostly in the West Webburn valley, but also in Bellever Plantation, at Archerton, and in Postbridge itself (Dare and Hamilton 1968). By the 1990s this number had risen to 10 to 15 breeding pairs, mostly in old deciduous woodland with some in the older plantation areas (Dare 1996).

At the Harford Common Bird Census site one pair has been regular yearly from 1987 to 2000, and it was joined by a second pair in 1998 and 2000. Up to four territories have been found most years at Burrator. At Huccaby two or three singing males were located in 2000, and an estimated five territories were occupied in 2001 (Hibbert 2001). During the fieldwork for the Devon Atlas Project in 2002, four singing males were found in part of the area around Huccaby, and a further two in the West Dart valley north of Hexworthy. Six territories were found that year in the Dart Valley woods below Bench Tor. Four were located in Hembury Woods, and two in the Bonehill area, and Yarner Wood. Single pair records are regular from most other woodland areas. Family parties of up to six are seen at times in late summer, and during the autumn and winter birds will regularly join foraging tit flocks.

Most birds are seen quietly searching for insects as they spiral up tree trunks, but occasionally they will search moss-covered boulders as at Burrator in March 1970, or even churchyard gravestones as at Okehampton in March 1976. A very unusual, and possibly unique, feeding activity was noted at South Brent on 14 March 2001 and Christow on 17 January 2002. In each instance individuals were visiting nut feeders. It was thought that the Christow record probably referred to the bird searching for insects on old nuts, rather than feeding on the nuts themselves, but these appear to be the first documented instances of this behaviour.

GOLDEN ORIOLE *Oriolus oriolus*
Rare passage migrant.
With only nine records this gaudy migrant is a rare sight on Dartmoor. As would be expected, all records have been from suitable woodland areas, and mostly during the spring passage peak of May to mid-June.

D'Urban and Mathew knew of one obtained in Okehampton Park, although no date is given. The dated records are as follows:
> 1928: Shaugh Prior, one in April.
> 1935: Lustleigh, 28 May, one.
> 1957: North Bovey, 17 April, one.
> 1969: Hembury Woods, 14 June, one.
> 1978: Fingle Bridge, 6 June, a male seen in flight.
> > Throwleigh/Gidleigh, 28 September, a male.

1990: Dartmeet, 5 May, a singing male at Badgers Holt.
1994: Piles Copse, 13 June, a female/immature male.

RED-BACKED SHRIKE *Lanius collurio*
Rare passage migrant. Has bred.

In the early summer of 1968 Dave Bubear was walking over South Down, above the Okement Valley. His attention was drawn to a bird that flew up into a hawthorn bush, and he was astounded to find it was a male Red-backed Shrike. He returned the next day and found that a female was also present. They gave every indication of nesting in nearby deep thicket. Through careful and quiet observation over the next few weeks he was able to confirm that they had bred successfully, raising three young. This would prove to be the last known nesting pair in Devon.

The birds returned again in 1969 and 1970, being successful again, and fledging three and four young respectively. By 1971 work had begun on the new Meldon Reservoir, and although the pair returned, disturbance close to the nesting site was too much, and no nest was found or young seen. The female returned in 1972, but left again in the early part of the season. No birds were seen in 1973. Dave Bubear at the time kept the details of this breeding pair quiet for obvious reasons, but later when the details were made known, another record indicated that the pair may have bred prior to 1968. On the 7 July 1966 a party of six birds had been seen in the same general area by Dr. Malcolm Spooner (D. Bubear *pers. comm.* 2003, Sitters 1974.).

Although not giving any specific information on the Dartmoor region, D'Urban and Mathew considered the Red-backed Shrike a summer migrant in Devon, and not very common. At about the time that they were writing, in the late nineteenth century, a decline in numbers was being noted nationally. This decline began taking hold in Devon in the early 1900s. On Dartmoor, where numbers must always have been relatively low due to limited suitable habitat, this reduction in numbers meant that the bird's fortunes were never going to improve.

In the last 50 years migrants have appeared at 11 sites, but one stands out from the rest, having records in six years. This is the area of largely heather moor and mining gullies, centred on Vitifer and Soussons Down, and reaching as far as Grimspound in the east and Warren House Inn in the west. The earliest records here consisted of a male on 10 June 1950, and a pair the same year chasing bees on 14 July. Breeding, however, was not suspected. On 28 May 1973 a singing male was present, but was not seen again. Two birds were seen on nearby Challacombe Common on 3 July 1987, and a singing male was again present in the area on 27 to 29 May 1989. A female was seen on 18 June the same year. A male was reported on 26 May 1997, and the most recent record is of a female near Headland Warren Farm on 9 June 2001. This area obviously has always been very attractive to the species, and given the number of records, it must be quite possible that a breeding attempt was made at sometime in the past, and gone unnoticed. This could have been especially possible in the middle decades of the last century, before human disturbance had reached its present peak.

Records have been far more sporadic from elsewhere. These have mostly been obvious passage birds, stopping to feed before moving on, although a male near Tavistock on 27 July 1954, and a male, with a probable female present, just outside the town near Anderton on 7 June 1957 (H. Kendall *pers. comm.* 2004) could give rise to speculation that breeding may have been attempted. Males have been seen at Fernworthy on 25 May 1954, Throwleigh on 6 June 1985, Deancombe (Burrator) on 9 May 1994, Hemsworthy Gate on 27 May 1997, and near Ingra Tor on 7 June 2004. Females have been reported from Meldon Reservoir on 3 July 1990, near Powder Mills on 14 May 1998, and at Buckfastleigh on 3 June 1998.

Most Dartmoor records have been in the peak passage period time of late May to early June, with others in July suggesting wandering non-breeders. There has only been one autumn record, and that was of a juvenile at Cadover Bridge from 18 to 20 September 1996.

GREAT GREY SHRIKE *L. excubitor*
Winter visitor and passage migrant.
This visitor from Fennoscandia crosses the North Sea and arrives on the east coast of Britain from late September each year. After feeding in temporary territories for a few days, most birds move inland to find suitable winter quarters, or continue their migration (Wernham 2002). On Dartmoor there have been a few October records, and the bird at Fernworthy on 18 October 1992 would appear to be the earliest. Normally birds tend to filter westwards rather more slowly, and arrive at sites during November. They then may stay a few days, or take up a territory for the winter. If an individual winters, its territory can be restricted to a relatively small area, or it may range widely, depending on the abundance of food.

D'Urban and Mathew list one seen between Lydford and Bridestowe on 15 November 1876, one obtained at Ashburton with no date given, and one shot at Roborough in November 1849, which could have been on the edge of the Down. There appears to be a total lack of records from the moor during the early twentieth century, and the negative run was only broken in 1960 when a bird was seen near Chagford on 3 January. A spring migrant stayed on Cator Common, near Runnage, from 15 to 19 April 1963, and from then on records have been virtually annual.

Birds have occurred at 26 sites around the moor, usually in areas of grass or heather moor with numerous hawthorn bushes for lookout posts. Latterly they have also been found in clear-fell or newly planted conifer areas. One area is traditional, and has had either wintering birds or spring migrants in 25 years out of the 37 since 1967. This is the area of valley favoured also by Red-backed Shrikes, centred on Vitifer. Depending on food availability, birds will range from here to Birch Tor, Soussons Down and

Challacombe Down, perhaps at times as far as Cator Common or even Bellever. During 10 of the above mentioned years, birds have stayed from early winter through to spring. Other years may well have had birds staying the winter, but records are more sporadic and the picture is less clear. Spring passage migrants have been found in eight years using the area from mid-March to late April. The earliest bird in the autumn here was on 20 October 1974, and the latest in the spring on 26 April 1977. They are usually seen singly, but up to three were present in January and February 1975. With the national decline of this bird as a wintering species, records at Vitifer have not been so regular in recent years.

Other areas have also proved attractive over the years, even if to a much lesser extent than Vitifer. Roborough Down and Horrabridge Common had records in 1969, 1971-72 and 1982-83. The 1970s records concerned a wintering individual, and the others were of shorter duration. The Haytor/Buckland Common area had records, usually single day sightings, in 1973, 1974, 1980, 1982, 1994, 1998 and 1999. Two migrants were seen on the late date of 29 April 1994. Hennock Reservoirs attracted birds on 1976, 1977 and 1979, and birds stayed the winter in 1980, 1993 and 1994. Fernworthy, although not having many records, has had autumn birds in 1981 and 1992. One stayed the winter of 2003/2004. Mostly one day records, of migrants or birds searching for suitable

Great grey shrike

199

habitat have come from Bellever, Belstone Cleave, Okehampton, Manaton, Sourton, Ashburton, Cator Common, Beetor, Spitchwick, Meldon Reservoir, Bowermans Nose, Hingston Down, King Tor (Walkham), and Whiteworks, where the latest bird in spring on Dartmoor was seen on 8 May 1982.

JAY *Garrulus glandarius*
Resident breeder, and occasional irruptive passage migrant.

The Jay is a woodland bird that is normally very sedentary, and seldom strays from its chosen habitat. On Dartmoor the oakwoods of the river valleys are its traditional home, but it has spread to the lower conifer plantations, although it does not appear numerous in some of the higher ones. None were encountered in Malcolm Spooner's survey of the higher plantations in 1937. In the Postbridge district, the first pair was found in Bellever plantation in 1957. Between one and five pairs were later located in the 1950s and 1960s, mainly in Bellever plantation, but with a pair in the West Webburn valley (Dare and Hamilton 1967). By the early 1990s, although numbers had increased, the increase was not substantial and only five to 10 pairs were estimated to be breeding. The increase was largely due to the colonisation of the matured conifer plantations (Dare 1996).

The more mixed woodland around Burrator, although smaller than the study area at Postbridge, has had at least six pairs most years since 1996. The altitude, of course, is much lower at Burrator. The Common Bird Census site at Harford has regularly held one or two pairs between 1987 and 2000, but three pairs were on territory in 1992. Jays are not normally seen outside of woodland, but occasionally they will make trips into more open habitat, no doubt quietly exploring for food. In 1990 a pair was watched foraging in the valley at Coombeshead, Burrator on 6 May well away from any woodland, and on 8 August 2000 two were moving between hawthorns on open moor near Venford Reservoir.

In the autumn birds become very active in their search for acorns, and obsessed with burying them as a future food source. They become so reliant on the acorn crop at this time of year, and in the winter ahead, that if it fails this normally sedentary bird is forced to move out of its home range completely. This is a very rare occurrence in our region, and an exodus of local birds in any numbers has never been noted, but some movements of Jays in autumn have been seen. The origin of the birds in these cases is uncertain, although the sight of the passage has been quite spectacular. At Sticklepath on 18 and 19 October 1930, numbers were seen moving northwest in parties of from nine to about 55. Some flew high, whilst others were so low that they had to rise to avoid hedges and trees. Some were seen to perch in trees for a few moments, as if tired, before resuming their flight. In 1957 a migratory influx was noted in Devon from late September. On Dartmoor the only instance of this influx was at Wrangaton, where six flew east and three south at the height of about 200 feet. on 5 October.

It is not known if these two observations involved a local or perhaps national movement, but the next movement was very different and involved birds known to be from much further afield. In 1983 the English and continental acorn crop failed, and an invasion of hundreds of birds was noted along much of the south coast. There were indications that more local populations were moving ahead of the continental influx which took place in October. Eighteen at Fernworthy on 25 August was an unusually high number there, but was far surpassed by the numbers in mid-to late October. On 11 October between 75 and 100 Jays moved through the East Okement valley between 11 am and 12.48 pm. At the almost treeless area of Whiteworks, 15 were present about 19 October, and 30 were at Belstone on 24 October. Flocks moved through quite quickly in their search for food, but there were records of birds returning north in the following spring. The only Dartmoor report, though, was of 24 at Merrivale on 6 May 1984.

MAGPIE *Pica pica*
Resident breeder.
Since the reduction of keepering on country estates during and following the First World War, this species has increased greatly in numbers. This increase culminated in the expansion into more and more suburban and urban areas over the last 30 years or so. The population is now generally thought to have stabilised at a very high level.

On Dartmoor, although numbers have undoubtedly increased over the years, the population swing has not been all one way and at certain locations a decrease has been noted, probably due to habitat change. Birds were found above the highest levels of cultivation in 1937, and during the Postbridge survey years of 1956 to 1967 there was estimated to be a population of about 50 to 75 pairs, with birds frequently nesting in small hawthorns on open hillsides (Dare and Hamilton 1968). However, by the early 1990s, although still common, the population had mainly shifted away from farmland to the edges of the plantations, and the total number of breeding pairs, although not counted, was thought to be less than the 1956-67 total. The decrease was thought to be due to loss of high thorn hedgerows in some localities (Dare 1996). Even in times of expansion there can be yearly fluctuations in numbers, as at Moorhaven Hospital near Bittaford, where at least 30 pairs were found in the area in 1986, but fewer than 20 in the same area the following year. At the Harford Common Bird Census site numbers have stayed constant, with six or seven pairs most years 1988 to 2000, although eight were found in 1988 and 1998, and nine in 1995. The Dartmoor Moorland Breeding Bird Survey of 2000 found Magpies present in moorland fringe areas containing scrub and woodland edge, and often associated with conifer plantations. They were also found to be common in the Princetown and Two Bridges areas. The survey estimated 192 +/- 56 territories as being the total for the moor (Geary 2000). The most recent fieldwork for the Devon Atlas Project in 2002 found eight territories in the Bonehill area tetrad, with two at Hexworthy and Yennadon Down tetrads.

Birds are highly sedentary and hardly ever move far from their breeding territories, which are normally close to natal areas, but at times birds can be found in areas further into open moor than usual. On 17 January 1993 a bird was at Lydford Tor, and the same year one was at Evilcombe in the Plym valley on 1 March. A pair were at Rowtor on 18 April 1997. Although usually solitary or in pairs, Magpies will flock in autumn and winter. This is especially noted in the cold weather of December to February when a good food source will attract small flocks of up to 20 or so, and occasionally more. At Yelverton on 23 January 1977, 31 birds plus the resident pair were counted, and at Ilsington up to 30 were present in a field daily spread with slurry during December 1985. The highest ever number was noted at nearby Liverton on 27 March 1985, when 35 were counted. This is a very late date for a flock of this size, most birds being paired and in territories by then. Other large gatherings in the winter have included at least 24 at Hennock Reservoir on 30 January 1981, and 30 at Dousland on 20 February 1988. Autumn gatherings have included 27 going to roost in woods bordering the Lee Moor China Clay complex on 13 September 2001, and 26 in the same area at Portworthy Dam on 19 September 1995.

NUTCRACKER *Nucifraga caryocatactes*
Very rare vagrant.
1968 will always be remembered ornithologically as the year of the Nutcracker invasion. At least 200 arrived in East Anglia from August onwards, and these very confiding birds spread westwards, reaching most parts of the country. Dartmoor would have missed out were it not for a single bird that was seen well at Moorgate, Wrangaton on 22 September.

D'Urban and Mathew mention that the species is included in Dr. Tucker's *List of Birds discovered in the Parish of Ashburton*, but there are no further details to back up the record, and it should be considered doubtful.

JACKDAW *Corvus monedula*
Resident breeder.
Being colonial nesters and social birds at all times of the year, Jackdaws are frequently heard and seen around moorland towns as well as in more rural areas. They co-exist with mankind very well and there is no evidence to suggest that they are anything other than a successful species with a stable population.

D'Urban and Mathew referred to them as being numerous on the Dewerstone, as they are still today, nesting in the cracks and fissures of that rocky outcrop. They also note birds as nesting in rabbit burrows at Vitifer. By the 1950s they were still there, but mainly nesting in the ruins of the now defunct mine workings. Some were still nesting there in the early 1990s. In the Postbridge area generally the total from the 1950s and 1960s of about 40/50 pairs still held good in the early 1990s (Dare 1996). The ruined buildings around Powder Mills held 20/25 pairs in the 1950s and 1960s

(Dare and Hamilton 1968), and 13 pairs were still there in 1999. They are the ever-present threat to open chimneys in most towns and villages, and these convenient man-made nesting sites can hold a considerable population in some localities. About 40 pairs of chimney nesting birds were estimated in Ashburton in 1992, and at least 20 pairs in Buckfastleigh in 1995. Five pairs nested in chimneys at Huccaby in 2000 and 2001. Even in out of the way places, chimneys are still an attraction, as with the two prospecting the chimney of Nuns Cross Farm in April 2000, and the two pairs in the ruins of Longbettor in June the same year. At Harford between five and 10 pairs nested annually in the farmland Common Bird Census site between 1987 and 2000, with an average of eight. Tree hole nesting is not very well recorded, although it must be far from uncommon. One or two pairs bred in tree holes at Yarner Wood in 1988, 1989 and 1994, and at Burrator where birds had been noted tree hole nesting in the 1970s, a bird was feeding young in an old woodpecker's hole in 1981, and the same trees are still inhabited each year. The sedentary nature of local birds was confirmed when one was found dead at Okehampton in 1990, that had been ringed there as a chick eight years earlier.

Birds will flock and roost together from mid-summer, with numbers increasing in autumn and winter when they will flock with Rooks and other corvids. The largest mid-summer flocks noted have been at least 300 at Bench Tor on 24 June 1979, and the same number at the Dewerstone on 22 June of the same year. These summer flocks often form to take advantage of the abundance of defoliating caterpillers in the oak woods at this time of year. By autumn they can be found on stubbles, as the 200 at Bridford on 28 October 1999, but the big numbers in winter are usually connected with communal roost sites. Flocks in the low hundreds at this time of year are not too unusual, but the largest congregations have been about 1000 with Rooks at Hennock on 15 November 1980, and a huge one to two thousand roosting in trees at Burrator at a time of severe weather on 3 February 1962.

ROOK *C. frugilegus*
Resident breeder.
Rookeries on Dartmoor are mainly small. Some are traditional, and have been established for many years, whilst others, usually the very small ones, come and go at the whim of the few inhabitants. Interference by man, either direct through destruction of birds or nests, or indirect through changes of habitat, can of course also have an effect on a rookery's fortunes.

Recording of rookeries has never been thoroughly carried out in the region. The national 1996 BTO Sample Rook Survey, only covered certain areas picked at random by computer. The results as far as Dartmoor was concerned were interesting, but of necessity not very complete. In the last 70 years or so rookeries have been recorded in at least 40 locations, although with poor overall coverage it is likely that the actual total is considerably higher. Few rookeries have been counted over any period of years, and so population shifts, etc., are rarely available. Where counts have been made consecutively, the bird's fortunes have showed some variance. At Dunstone Farm, near Cadover Bridge, there has been a consistent rise in the number of nests from 21 in 1990 to 55 in 2002. This has then to be one of the largest rookeries, certainly on western Dartmoor, although on the eastern side of the National Park larger rookeries have been noted at Christow and Dunsford, with both having about 120 nests in 1995. During the 1996 BTO Sample Rook Survey, an area around North Bovey contained three rookeries, totalling 97 nests, and two rookery sites at Moretonhampstead had 49 nests. A large rookery in the central area of the moor has been at Prince Hall for many years. The number of nests, at two sites, in 1991 and 1996, was 40. Broadaford Farm, in the West Webburn Valley which held about 30 nests in the 1950s and 1960s., had the number increased to between 42 and 52 by 1992. Conversely some sites have shown a decline. The rookery by Harford church has dropped in numbers from 23 nests in 1988 to about half that figure by the mid 1990s, and shows no sign of recovering. The trees near Oakery Bridge, Princetown had a small rookery in the late 1980s and 18 nests were counted in 1990. However the rookery had disappeared totally by 1995, perhaps as a result of disturbance and the thinning out of many trees by winter gales. A few pairs returned in 1999, but not apparently since. The reasonably large site at Merrivale that had 35 nests in beech trees in 1994 and 1996 could only muster about 22 nests in 2002. It must be stressed that the increases and decreases, of which the above are just examples, are instances of local shifts in numbers only, and that the overall view must be of a stable population on and around Dartmoor.

After the young have fledged, post-breeding flocks form and numbers can be substantial. Some flocks venture far onto the moor to find suitable areas of short grassland habitat on which to feed. Their main food source at these locations at this time of year are the leather jacket larvae of the Crane Fly (*Tipulidae sp.*) Flocks will sometimes fly to small, obscure locations to feed in very restricted areas. An example of this was the flock of about 160 birds that flew up every day for about a fortnight to Little Links Tor from 30 May 1959, to feed amongst the heather. Larger numbers will frequently collect at traditional areas of grass moor. One area that gets flocks every year is the moor around Cox and Staple and Roos Tors, together with the Walkham Valley as far east as Great Mis Tor. Numbers can be so large here as to provoke speculations about how far do birds travel to feed on the moor. The flock size is too large to be made up of birds from local rookeries. Over 600 were in the area on 8 June 1999, but in 1991, a very prolific year for Crane Flies, over 1500 were feeding between Roos Tor and Great Mis Tor on 16 June. Gatherings of several hundreds are also regularly seen in areas of the Swincombe valley, near Wistmans Wood, in the Plym valley near Hartor Tors, Harford Moor, and Penn Moor.

There is evidence that birds leave the high moor in winter, even in mild winters (Dare and Hamilton 1968), and most of the large flocks seen at this time of year have been in the eastern, more agricultural portion of the National Park, such as the 300 at Bridford on 23 November 1996. Elsewhere, gatherings have usually consisted of birds going to roost, as with the 140 that roosted in the plantation at Fernworthy on 12 December 1988, or the 93 flying over Holne on 11 December 2001 heading towards Spitchwick/Leusdon at dusk.

CARRION CROW *Corvus corone*
Resident breeder.

This species is common at low density almost everywhere on the moor. It is only restricted as a breeder by the lack of trees and bushes on the highest tops. The Postbridge survey in the mid-twentieth century estimated about 100 to 150 pairs in the area, with the density especially high in the West Webburn Valley where it was thought they could have attained 20 breeding pairs per square mile (Dare and Hamilton 1968). A follow-up survey of the area in the early 1990s found fewer in the West Webburn Valley, possibly due to the loss of hedgerow trees on some farms (Dare 1996). Overall numbers were thought probably to be below the 1956-1967 levels, despite the utilisation of new coniferous plantation nest sites.

The Dartmoor Moorland Breeding Bird Survey of 2000 estimated a total of 785 +/- 148 territories for the whole moor. They were found to be widespread, and more common on fringes, near woodland or enclosures (Geary 2000). At the Harford Common Bird Census site, the average number of territories annually from 1987 to 2000 was 15, with peaks of 21 in 1993 and 1994. Although mostly sedentary, birds will move out of an area ahead of severe winter weather. This was noted at Wrangaton in November 1962, when a substantial drop in numbers occurred before deteriorating weather. These movements are perhaps very local, as this is a species that would surely gain from the misfortunes of other, more susceptible species, in times of freezing weather.

There appears always to be a nucleus of non-breeding birds about the moor. These form small flocks in April and May, the largest seen being 49 at appropriately Crow Tor on 5 May 2000. By late summer, flocks, now augmented by young birds, are seen around the moor in suitable feeding habitat. These flocks can be up to about 100 birds at times, but by far the largest number was 320 at Broad Down on 15 July 2002. Two days after this record number, at least 100 birds were roosting in Raddick Plantation at Burrator on 17 July. Winter flocking has been recorded to a rather lesser degree, with most records in the 20 to 50 range. Two flocks that greatly exceeded this, however, were the 200 or so at Owley around Christmas 1934, and the 280 at Cator Common on 10 November 1979.

HOODED CROW *C. cornix*
Very rare visitor.

Only four records of this wanderer from the North have been noted on Dartmoor. A bird was seen near Grimspound on 30 June 1969, and another at Haytor on 25 May 1974. One was watched at Meldon Reservoir on 29 November 1979, and the most recent record was of a bird on the moor near Okehampton on 4 November 1989.

Ravens

RAVEN *C. corax*
Resident breeder.

Nothing is more evocative of the wildness of the Dartmoor hills than a Raven flying high overhead on a bright spring day, calling to its mate. The deep, vibrant croaking calls completely compliment the wide vistas of the moor. The Raven must always have been a part of Dartmoor's natural history and folk lore. There is a Raven's Tor in the Bovey Valley above Lustleigh Cleave, and Crossing, in his 1905 *Guide to Dartmoor* lists four Raven Rocks around the moor. These are situated by Luckey Tor on the Dart, near Ausewell Rocks in Buckland in the Moor, by Leigh Tor on the Dart overlooking Holne Chase, and at West Down above the valley of the Tavy. In times past these were all probably nesting sites, and Pidsley, in the late nineteenth century commented that a pair used to breed at the West Down site.

As can be seen from the above, the traditional nesting sites of Ravens have always been high, rocky vantage points, either overlooking a river valley or on a remote tor. D'Urban and Mathew mention pairs at the Dewerstone and Tavy Cleave, that a pair bred on a tor near Lydford in 1883, and that young had been obtained from rocks near Haytor in 1890. Birds have always had some degree of persecution on the moor, mainly from farmers who feared for their livestock. This persecution, involving destruction of birds and nests, was noted in D'Urban and Mathew's day, and even then it was evident that certain sites were no longer safe for nesting birds. With the increase in tourism and moorland walking in the twentieth century, the persecution took on a new face in the form of disturbance. Ever more sites became unusable due to human pressure, and nesting on natural rocky outcrops and tors became largely a thing of the past. Traditional sites on tors almost disapeared in the last years of the 1900s, and the one or two sites where birds may just be hanging on are not identifiable for security reasons.

As natural sites declined quarry sites increased. There was a certain degree of security at these sites, and birds did rather well. Unnamed sites were noted in the 1930s, and a pair was breeding in an old quarry near Okehampton as early as the 1890s. Granite quarries at Haytor and Swelltor have had nesting pairs up until 1980 and 1998 respectively, and Whitecleaves and Bullacleaves quarries at Buckfastleigh have pairs annually. Birds have also bred on occasion at Merrivale, and Pitts Cleave Quarry, Tavistock. Recently the china clay pits at Lee Moor have been used. Five pairs bred there in 1995 and six the next year. Ever the opportunists, Ravens have nested in a few more unusual sites in the past. For at least 30 years up until 1966 a pair nested on the top of an old railway viaduct pillar at Cornwood, and throughout the later 1950s and 1960s a pair nested in the mining gullies at Vitifer (Dare and Hamilton 1968). The site was abandoned in 1969 after a rock fall.

The acceptance of trees for nesting has saved the breeding population on Dartmoor from total decline. This habit was first noted in the 1920s, and increased greatly by the 1940s. It was during this period that an increase in sightings was noted in many areas around the moor. The maturing of plantations was a great help in providing suitable sites, but pairs in the early days were noted in a variety of trees, including Scots pine, sycamore and ash. Plantation nesting in spruce was first noted in 1936 in a small plantation on the western slopes of the moor, quite possibly Baggator. A pair nested yearly at Archerton from the mid-1950s until 1990, when the wood was blown down in a January storm. The Common Bird Census site at Harford had birds present most years from 1987 to 2000, but territory was only held in 1988, 1990, and 2000. A pair bred in trees at Owley Bottom in 2001, the first time in at least 60 years. Today, with the population buoyant, pairs breed in suitable trees, and even a few not so suitable, virtually anywhere around the moor edge and in the interior.

Ravens are very early breeders. Pairs regularly have eggs by late February, and this leaves them at risk from severe late winter weather. In the bad winter of 1947 at least four rock eyries were not tenanted, due to the nests being deep in snow when incubation should have begun. Tree nests appeared to do a somewhat better that winter, although broods a little late. After fledging, birds will wander, usually in a family group. Families usually stay together for much of the summer, but by late summer or early autumn the bonds relax and the juveniles begin to make their own way in life. It is during this time that gatherings of Ravens on the moor have reached their highest numbers, no doubt augmented by birds of the year. At least 100, roosting in tall larch trees in the Glaze Brook Valley on 30 August 1976, was the largest single gathering recorded.

Ravens congregate for two reasons; to roost communally and to take full advantage of any new food source as it becomes available. Gatherings for either reason can take place at any time of the year. It is thought that paired Ravens stay in their territories throughout the year, and that the assemblies comprise mainly immature birds and non-breeders, many of whom, even if paired, have not been able to obtain a suitable nest site (Ratcliffe 1997). The roost of up to at least 80 birds at Wrangaton and Piles Copse between the third week of January and Easter 1955, when birds should normally have been breeding, supports this. Many of the roosting sites remain unknown, with the only indication of their whereabouts being the flight lines of birds morning and evening. In 1973 and again in 1979 between 70 and 80 birds were regularly noted in the evening heading north off the moor at Prewley. The area around Princetown has also witnessed some big movements, probably linked with a corvid roost in the area at either Beardown Plantation or Holming Beam. The highest numbers reported have been 73 on 9 September 1990, and a single flock of at least 50 flying over on 6 September 1996.

The tors forming the western escarpment of the moor from Meldon Reservoir south to Tavy Cleave, Cox Tor and the Ingra Tor area, attract many parties of Ravens throughout the year. These groups appear to wander rather aimlessly over the ridges and hills, coming together to form larger gatherings then splitting up again, sometimes within a matter of minutes. An instance of this was noted over Tavy Cleave on 13 June 1999, when 28 birds came together, soared for some minutes, then split up with pairs and small groups drifting off in several different directions. A very similar occurrence with 43 birds involved was witnessed at nearby Standon Hill on 9 July 2002. The search for carrion is undoubtedly the spur for these wanderings, as with the 63 found feeding on two dead sheep at Meldon Reservoir on 25 July 1973.

STARLING *Sturnus vulgaris*
Resident breeder, passage migrant and winter visitor.
The Starling has declined markedly as both a breeding bird and a winter visitor in Britain over the last 20 years. The reason for this is unclear, but is probably agriculture related. On Dartmoor breeding records are few, so little comparison can be made with times past. However, writing in the early 1830s the Rev. Thomas Johnes made the interesting comment that Starlings did not breed in the neighbourhood of Dartmoor. They arrived in September, were found in vast numbers in the fields in December and January, but left again in late January or February. In the 1890s, D'Urban and Mathew made a similar comment that the Starling was only an autumn and winter visitor to the moor. During the first half of the twentieth century little comment was made, but by the 1950s there was a breeding population on the moor, and in the Postbridge area the numbers increased from just one or two pairs during 1956/1957 to 20 to 30 by 1966/1967. However numbers fell dramatically to only about five pairs by 1990-1995. Most nested in farms and houses in Postbridge itself (Dare 1996). At the Harford Common Bird Census site two pairs bred in 1990, but this reduced to one in 1992, and continued with one until 1995. In 1996 and 1997 birds were present but no territory was held, yet in 1998 12 territories were held. This proved to

be a one off, and by the next year no birds bred. None were present in 2000. It can be deduced from casual records that although Starlings breed on Dartmoor in areas where there are buildings for nest sites and grassland for feeding, they are not to be found in large numbers, and must at best be thought uncommon. Although not a target species for the survey, the Dartmoor Moorland Breeding Bird Survey of 2000 found the species very scarce.

Starlings have the distinction of possibly being the only British passerine species where the juveniles flock together, away from most adults after fledging. This phenomenon has been noted on Dartmoor on some occasions. Congregations of juveniles, no doubt birds reared in areas off the moor, have been noted feeding on grass moorland on several occasions in mid-summer. Many of these flocks have been small, but others have been quite large, suggesting that individuals must have flown up onto the moor from some distance. In 1987 a flock of about 600, mainly juveniles, was noted at Bowermans Nose on 26 June, and in 1979 a larger flock of about 1500, again mostly juveniles was recorded at Birch Tor on 24 June (Smaldon 1993).

Although juveniles wander, the adult population of Starlings in Britain is thought to be largely sedentary. In late October to early November the resident population is augmented by thousands of birds arriving on British shores from northwestern Europe. These continental flocks first arrive from over the North Sea to the East Coast, but very quickly move on to find winter quarters. In our area the arrival of these continental birds is not well documented, but 1000 at Merrivale on 7 November 2000, and the same number in fields at Princetown two days later, were no doubt continental immigrants. Where the build up of birds does show during the winter is in the communal roosts. There have been many in the Dartmoor region over the years, some of long duration, and some used for only a year or two. The main roosts have been as follows:

Princetown: A medium-sized roost was noted here in a conifer plantation by the railway station during the winter of 1935/1936. There were then no records until 2000 when the thousands of birds flying west up the West Dart at Huccaby on winter evenings were found to be roosting in the small plantation at Princetown, just south of North Hessary Tor (Hibbert 2000). This is almost certainly not the same plantation used 65 years earlier, but it does appear to occupy a very similar location. This small wood is very exposed and must be amongst the highest on Dartmoor. Why thousands congregate here, perhaps from as far away as Ashburton, with others flying up the Meavy and Walkham valleys can only be guessed at. Birds have also been noted approaching from the Merripit Hill, Brimpts Plantation and Postbridge directions. The roost was still being used in 2002, although possibly not after that. There were signs in November 2001 that an alternative roost elsewhere was being used. After the normal build up, numbers dropped dramatically in the latter half of November (G. Easton *pers. comm.* 2002). The same pattern was noted in the second winter period of 2002.

South Brent: A roost was recorded at Whiteoxen in 1954, and was found to replace the Slapton reedbed roost in late winter. The same pattern was noted in 1955.

Soussons Plantation: A large roost was first noted here in 1963. This was a vast roost that drew birds from as far afield as East and North Devon and eastern Cornwall. By 1974 it was considered the major roost in Devon from December onwards, and that year an estimated 700,000 went to roost on 24 December. It was abandoned, no doubt much to the forester's joy, by 1978.

Lee Moor: At the wood by Lee Moor House an estimated 50,000 roosted on 16 January 1966.

Fernworthy: On 15 November 1972 an estimated 100,000 Starlings roosted in the plantation. Perhaps these were birds that had been disturbed from Soussons.

East Okement Valley: An estimated 200,000 birds roosted in a plantation at East Okement Farm on 29 January 1991. The size of this roost suggests that it must have been a regular site, for a time at least, but there were no records before or after.

Thorndon Cross: Although to the west of Okehampton and outside the National Park, this has been by far the biggest roost in the Dartmoor area since 1992. In 1995 birds were followed to the roost from as far as the outskirts of Exeter, and it was estimated that about a million were roosting in January and February, and a million and a half during November and December. Up to two million roosted in November and December 1996, and the peak figure has remained about the same each winter since. Flocks can be seen on many winter evenings, over-flying the moor from many directions, heading towards this roost.

Other roosting sites have been used by smaller numbers at Cornwood in 1937, Grenofen in 1955, Bellever in 1957, Black Tor Copse in 1969, Whiddon Down in 1982, and Okehampton in 1983.

By March, roosting sites are being abandoned, and the continental wintering birds are making their way back east. Large daytime flocks are seen feeding avidly on short cropped grass moorland at this time, before departing eastwards. A feeding flock of 20,000 was seen in the Dunsford area on 14 March 1995, and on the same date in 1998 two feeding flocks, totalling at least 25,000, were on small areas of grass moor near the Devil's Elbow, Princetown. Birds in both these records were not present the next day. Five thousand were seen near Two Bridges on 9 March 1991, and 2000 on cattle pasture by Jay's Grave on 5 March 1990.

Two interesting ringing records have shown where at least some of our wintering birds come from. A bird ringed at Kanningrad, in the former USSR, on 20 October 1982, was found dead in Okehampton on 23 March 1983, and a bird ringed at Overtjssel in the Netherlands on 3 February 1996, was re-trapped at Okehampton on 19 November 1996. This was presumably a bird from the Baltic, that spent the winter in the Netherlands one year, and in Britain the next.

ROSE-COLOURED STARLING *S. roseus*
Very rare vagrant.

In recent years incidents of this species occurring in Britain have increased greatly. Arrivals largely conform to two patterns. Adults arrive in June and July, with juveniles in the autumn. Devon has had its share of this increase, and Dartmoor has received its first two acceptable records. The first was an adult seen amongst the china clay workings at Lee Moor on 14 June 2001, and the second was a juvenile that was present at South Brent from 27 October to 3 November 2002.

HOUSE SPARROW *Passer domesticus*
Resident breeder.

Unfortunately, this is a species that everyone takes for granted. It is poorly recorded now, and was poorly recorded in the past. Although probably always present in moorland towns and villages, as well as more remote farmsteads, records are so few that the status of the population is almost impossible to assess. What records there are suggest a rather patchy distribution, with numbers increasing at some sites and decreasing at others. Its presence in days past was taken to be the norm at suitable localities, so two records published in the yearly *Reports of the Devon Bird Watching And Preservation Society*, are interesting in that they refer to its absence. In 1941 it was reported as entirely absent from the environs of the Holne Chase Hotel, in spite of the fact that there were occupied stables, a number of fowls, and any quantity of apparently suitable nesting sites in the area. In 1949 the species was found to be very local in the Chagford area, and absent from many farms except as an occasional visitor. The pair that arrived at Moorgate, Wrangaton on 16 February 1957, was the first there for 20 years. They stayed for several months but then disappeared without nesting. A similar situation occurred there with a pair in 1960.

Around Postbridge, about 25 to 50 pairs were estimated as breeding in the 1950s and 1960s. They were resident in small numbers about some farms, but not all, and the main centres of population were Bellever and Postbridge itself (Dare and Hamilton 1968). By the early 1990s a reduction to about 10 pairs was noted (Dare 1996). The reasons for this are unknown, but there has been a national decline in numbers over the last 30 years. The Harford Common Bird Census site had a single territory in 1992 and 1993. This then increased from two in 1994 to six territories in 2000. At Thimble Hall, Hexworthy two pairs breeding in 1999 had increased to five pairs the following year. Recent fieldwork for the Devon Atlas Project in 2002 found 21 pairs in the Bonehill tetrad, and eight pairs in the Yennadon Down tetrad.

There are few records of autumn flocking in the Dartmoor area, but 40 were present in Okehampton on 14 August 2001, and 45 on 24 October 2002.

TREE SPARROW *P. montanus*
Very rare winter visitor and possible passage migrant.

There have only been four confirmed records of Tree Sparrow from the Dartmoor region. Three of these were just outside the National Park boundary, but are worth noting here. Intriguingly, although two of them were nearly 20 years apart, they were from the same general locality. In 1953 three birds frequented the area of Brook House, Tavistock from the end of October until 9 November. In 1956 two birds were seen in the same place in January and February, and in 1975 one was seen in Tavistock on 10 February. This is officially the end of the Tavistock records, but in the late 1980s or early 1990s there was an unconfirmed report of a bird or birds near Longford Farm, east of Tavistock. The most recent record, and one still on the western side of the moor, was of two at Okehampton on 31 January 2001.

Tree Sparrows are not known as long distance migrants, although they do disperse further than the almost completely sedentary House Sparrow. In all probability the above records refer to birds moving westwards in winter from counties to the east where they breed, but there is just a possibility that a pair or two may have become resident somewhere on the western side of the moor, and gone unnoticed among their common cousins.

CHAFFINCH *Fringilla coelebs*
Resident breeder, passage migrant and winter visitor.
The Chaffinch is, and probably always has been, a common breeding bird on Dartmoor. Although the wooded fringes of the moor are its traditional habitat, it has had its population greatly increased by the acceptance of territories in the maturing conifer plantations. It was found to be a widespread breeder in the Postbridge area during 1956-1967, with a population estimated to be between 250 and 500 pairs (Dare and Hamilton 1968). In the early 1990s the population was thought to be still within those limits, although more conifer plantation had been colonised. At least 40 singing males were found in a part of Soussons in 1992, an area of plantation where the trees would have been too young for many pairs in the earlier survey years (Dare 1996).

At the Harford Common Bird Census site numbers rose from 31 territories in 1987 to 44 in 1991. Then came a big reduction in 1992, with only 29 territories. Territories then increased only slowly until a figure of 34 was reached in 2000, when the Common Bird Census ended. At Dunsford Wood Nature Reserve the picture was rather different. 1992 was a bumper year, with 26 territories the highest since 1980. The figure had reduced to 21 in 1993, and was between 14 and 19 in the years up to 2002. The 2000 Dartmoor Moorland Breeding Bird Survey found Chaffinches very common in moorland fringe areas near woodland and also in the plantations, and the population was estimated to be 1050 +/- 332 territories (Geary 2000). Fieldwork for the Devon Atlas Project in 2002 revealed 27 singing males/pairs in the Bonehill area tetrad, 13 in the Hexworthy tetrad, and about 100 in Holne Chase, extrapolated from counts in a small area.

In the autumn, local numbers are swelled by the arrival of continental immigrants from Scandinavia. These birds usually arrive in late October or early November, and at times visible migration is noted, as with the small parties flying north over Wrangaton on 22 October 1950, and the 110 moving west in the same area at Western Beacon on 7 November 2002. At Holne on 17 November 2002, an almost constant movement north was noted during the day. How long the flocks stay in an area depends greatly on food availability, particularly beech mast. In good beech mast years large flocks will stay well into the winter, or until the mast crop is exhausted. In poor years flocking will hardly be noticed. Three hundred were at Lydford on 1 November 1974, at least 400 were at Okehampton on 30 October 1989, and 250 at Bridford on 17 November 1999. If a good food source is located, flock size can build during the winter months, with the highest numbers for the region often being found during the December to February period. A flock of over 1000 was noted at Okehampton on 25 December 1970, and remained into January 1971. A flock in the Brisworthy area in 1979 increased from 250 on 10 January to about 600 on 4 February, with 500 still present on 4 March. Up to 1500 were in the West Webburn Valley on 17 February 1957, and the same number was present throughout the latter half of December 1983 at Sampford Spiney. However, the largest single count on Dartmoor was the 2250, counted in to roost at Soussons Plantation on 12 January 1983. Flocks disperse during March, and congregations of 100 at Cornwood on 24 and Soussons on 26 March 1996 were somewhat unusual.

BRAMBLING *F. montifringilla*
Winter visitor and passage migrant.

The first Bramblings to arrive in autumn from their Fennoscandian breeding grounds are usually seen in October. The very earliest records consist exclusively of birds heard calling as they pass over at night, thus showing that, at least in some autumns, there is a passage over Dartmoor before the first grounded migrants are seen. At Wrangaton the first birds were noted overhead on 1 October 1968, 2 October 1955, 4 October 1972 and 11 October 1966. Birds found feeding off passage are seen from late October and particularly in early November. Whether these birds stay to form the nucleus of larger wintering flocks or pass quickly out of the area is unclear. It could well depend on the abundance of beech mast, their main autumn and winter food source. Larger flocks at this time of year have included 40 at Holne on 7 November 1965, 41 at Horrabridge on 9 November 1972, and 80 at Postbridge on 7 November 1995.

As winter approaches in years of a plentiful beech mast, birds can appear quite plentiful beneath trees, frequently with Chaffinches. However, these feeding flocks are transient and will move out completely once the beech mast is depleted. In poor beech mast years Bramblings are difficult to locate. Even in years when they are quite plentiful, a large size flock would rarely be more than about 50 birds. Exceptionally flocks can be larger, as with the 150 in the Postbridge area on 13 February 1957, and the similar number at Bridford on 29 to 31 December 1994. Good numbers will also join mixed finch flocks if a good food source, other than beech mast, has been located. In January and February 1979 up to 330 birds joined a huge finch flock in kale fields near Brisworthy Plantation, with at least 50 staying into early March.

By March the big flocks are breaking up, with birds beginning their migration back to the Continent. By the end of the month only small groups at best are normally still around. At times a larger number will be seen, as with the 20 at Peek Hill on 20 March 1932, or the 40 at Soussons Down on 26 March 1996. Amazingly, a male at Soussons on 29 March 1997 was in full song. April sightings are unusual, but surprisingly there have been three sizable flocks reported during this month, suggesting perhaps that migrant flocks from elsewhere have come off passage for a time. A flock of 80 was near Grimspound on 4 April 1976, 29 were at Fernworthy on 5 April 2002, and over 100 were at South Hookner and West Combe Farm, North Bovey on 11 April 1970. The latest ever were four birds at Sticklepath on 17 April 1996.

An interesting ringing re-trap, confirming the movement of an individual across Britain during the course of one autumn/ winter, was of a male ringed at Spurn Point, Yorkshire on 26 October 1978 and re-trapped at Sampford Spiney, on 4 March 1979. Although not birds known for their winter site-fidelity, an adult male and an adult female ringed at Dousland on 27 March and 2 April 1994 respectively, were both re-trapped at the same location in January 1995.

GREENFINCH *Carduelis chloris*
Resident breeder, passage migrant and winter visitor.
Because of the changes in agriculture over the past half-century, this species has had to change from a bird of the countryside, to one that accepts gardens and parks in villages and towns. It is largely a lowlands breeder in Devon, and its distribution on Dartmoor tends to follow this new-found association with man in areas around the moorland fringe.

Breeding birds are decidedly scarce in many areas. In the Postbridge region it was thought in the 1950s and 1960s to be an irregular visitor in very small numbers. Its occurrence was usually associated with the winter, especially in cold weather. There had been only three summer records, and breeding was never suspected (Dare and Hamilton 1968). By the early 1990s the position had changed very little, although it was thought to have bred on occasions (Dare 1996). In more recent years breeding has been noted at Bellever. Birds have been noted at Princetown, despite its high and exposed location, during recent breeding seasons. At the Harford Common Bird Census site Greenfinches have never bred, and indeed between 1987 and 2000 they were only present as non-territory holders in 1995. The situation in the West Dart valley is rather complex and not yet fully understood, but birds bred at Huccaby in 2001, almost certainly breed regularly at Hexworthy, and are well represented in the Dartmeet area, where juveniles were seen in September 2001. Fieldwork for the Devon Atlas Project found four territories in the Bonehill tetrad in 2002, with seven in the Yennadon Down tetrad the same year.

Greenfinches are thought to be largely sedentary, with few going more than 20 kilometers from their natal area, but after the young have fledged family parties will form flocks. At least 70 birds at Shaugh Prior on 30 July 1989 were very early. At Bridford a ringing programme in 1995 and 1996 caught over 200 birds each year during August and September and with a re-trap rate of only about 10%. High local numbers were thought probable, and/or a transient population. In 1996 numbers were noted to have reduced rapidly when neighbouring stubble fields were ploughed. Movements of birds have been noted in late autumn, but it is not known whether these refer to a dispersing local population or migrants from further east or the Continent. In 2002 this movement, although very small, coincided with a much larger westward passage noted on the south Devon coast. At Wootton, Holne, a total of at least 36 birds passed over on seven dates between 11 and 29 November. Birds passed over going in many different directions, so it may have been more dispersal than true migration. Flocks feeding in winter are not noted so much these days, as they were 30 or so years ago, but 150 in the huge finch flock near Brisworthy Plantation in February 1979, the same number at Lower Cator on 24 December 1980, and 200 feeding on *brassica* heads at Chagford on 24 December 1990, are worthy of note.

Rhododendrons are often used by roosting Greenfinches, and at Kennick Reservoir a roost was noted between 1993 and at least 1996. Maximum numbers occurred during December to February, with the highest number being about 200 in December 1995. A much smaller roost was also noted in Rhododendrons at Burrator in the early 1980s.

GOLDFINCH *C. carduelis*
Resident breeder and passage migrant.

The Goldfinch is found around much of Dartmoor's periphery, wherever the habitat provides sufficient herbaceous food plants, and there are sufficient tall bushes and trees for nesting. It gets more uncommon in the interior valleys of the moor, and is absent from the high uplands. In the Postbridge area, from 1956 to 1967, only between one and five pairs were estimated to breed annually, and two to four of these pairs were in the West Webburn Valley. A single pair was usually at Postbridge village (Dare and Hamilton 1968). By the early 1990s the population had not increased, and was still thought to be about five pairs maximum, usually found close to habitations (Dare 1996). At Harford the Common Bird Census site had breeding pairs from 1989 to 2000. Up until 1993 two breeding pairs were normal. This increased to three in 1994, and had reached six by 1999. In 2001 three pairs probably bred at Huccaby House. During the 2002 Devon Atlas Project fieldwork, four territories were found in the Bonehill area tetrad, with two near Ditsworthy Warren, and one in the Yennadon Down area tetrad, and one at Sticklepath. Five birds were located along 2.5 kilometres of the Swincombe Valley on 4 May 2002. In general, it appears to be a species that can be found where sufficient suitable habitat exists, but it tends to be at a rather low density.

After the breeding season family parties can be seen and heard as they noisily make feeding flights within a limited area. As autumn progresses family parties form flocks and roam a larger area searching for food, especially the ripened seeds of the Thistle. From late August to October these flocks are widespread and can number over 100. The latter half of September is a particularly good time for these flocks, and the maximum count in recent years at this time was the 200 at Fernworthy on 24 September 2002. In recent autumns small flocks have been located feeding on dried *erica* and *calluna sp.* seeds, on the heather moor around Warren House Inn. It is probable that the autumn flocks contain migrants, but as yet this has not been verified.

215

In the winter Goldfinches are very scarce. By far the bulk of the breeding population makes an altitudinal migration, and comes off the moor for the cold months. It is likely that birds do not travel too far, many resorting to agricultural land, like the 200 that fed on, and below, *brassica* stumps at Chagford on 24 December 1990, and the increasing number that take advantage of garden feeders in moor edge towns and villages. Numbers present can vary in the winter from year to year, perhaps depending on the abundance of food or severity of the weather. In the well-watched area of Huccaby there were many sightings in the 2001 winter periods, as against almost none in 2000. At the end of the winter, with local birds returning to territories and perhaps migrants going north, some large gatherings occur on food plants not exploited earlier. At Hennock Reservoirs on 20 March 1977, at least 300 were watched feeding on Tamarack cones, and at Lydford Forest on 13 March the same year 200 were present, perhaps also taking advantage of a good cone crop.

SISKIN *C. spinus*
Migrant or resident breeder, passage migrant and winter visitor.
Up until the 1950s the Siskin was a winter visitor in very small numbers to the lower reaches of Dartmoor streams, feeding mainly in alder. Then in 1957 three birds were seen in Bellever Plantation on 15 April. Subsequently two pairs were located and one pair bred successfully, raising five young. Birds were also noted at Fernworthy in May of the same year. The colonisation of Dartmoor's conifer plantations had begun.

Siskins were noted in Soussons Plantation in the summer of 1958, as well as the two localities favoured the previous year. By 1960 birds had been seen in suitable breeding habitat at Tor Royal, and Postbridge, and in 1961 an individual was seen at Huccaby, although this perhaps was a false omen as the first confirmed breeding there was not until 2000. Burrator had its first record in May 1969, and although numbers increased slowly at the early sites, it was the late 1970s before breeders were suspected elsewhere. Birds were at Venford Reservoir in 1978, and in 1979 there were real signs of an increasing population with at least three pairs at Burrator on 2 June, and 15 birds in five areas of Fernworthy Plantation on the 3 June. No doubt Bellever and Soussons also saw increases about this time but coverage of these areas was by then not as complete as it had been in the 1950s and 1960s. By the 1980s there were signs of the breeding Siskins spreading to much smaller areas of conifers, with even areas of mixed or deciduous trees being used. This was a phenomenon that developed further in the 1990s. A small area of conifers at Dousland, and adjacent gardens, had birds in the summer of 1980, with proved breeding of at least one pair in 1984. Birds became regular in Yarner Wood in 1983 and subsequent years, with possible breeding in 1989, and the first confirmed breeding in 1991 (Page 1992). In 1997 a pair was watched in the oakwood of Piles Copse on 25 May. Display and song flights were noted. Even the small area of conifers around the Gutter Tor Scout Hut had potential breeders in the June of 1999.

By the 1990s Siskins were so widespread during the breeding season that the total yearly records were certainly underestimated. At Thornworthy birds were present in a larch copse by Richard Waller's garden throughout the 1990s. They bred most years, with an estimated four or five breeding pairs in 1993 and 1996. With this concentration in a relatively small area, and if the ratio of pairs to area held true for the larger sites, even the highest counts for the main plantations were potentially well below the actual numbers present. It has been estimated that 10 to 20 pairs bred annually in areas around Postbridge by the early 1990s (Dare 1996). About 40 birds were present in about half the plantation area surveyed at Soussons on 2 April 1997, including 11 singing males, although at this early date this total may have still represented migrants moving through as well as residents, and at least 10 singing males were counted at Bellever on 18 June 1996. Siskins were noted at over 30 additional summer sites between 1990 and 2002, where they had not been seen earlier. The species is without doubt still increasing, and the only note of pessimism must be that numbers could reduce with the clear felling of much of the main plantation areas.

Once the young have fledged, family parties will feed in the general area of the nesting site, but by August when most of the local food is exhausted birds disperse out of the area. At Thornworthy, with its well-observed annual breeding population, birds are always gone by the end of August (R. Waller *pers. comm.* 1996). Then follows a period when birds are scarce in the plantations. Where locally bred birds go is not known, but some British ringed birds have been found in winter in Iberia and North Africa (Wernham *et al* 2002). Forty at Fernworthy on 31 August 1990 was an unusually large number for the time of year. By September numbers are still normally low, with 110 at Fernworthy on 25 September 2002 being uncharacteristically high, and indicating an early arrival of birds from further north in Britain, or perhaps from the continent. During October birds from Scandinavia pour through Britain. This passage is noted regularly on the South Devon coast, and the last week of the month produces the largest autumn numbers seen on Dartmoor. Large concentrations have included at least 100 at Burrator on 21 October 1984, 125 there on 24 October 1993, and at least 80 at Fernworthy on 27 October 1989. The largest flock at this time of year, however, was 250 at Hennock Reservoirs on 25 October 1993, which was a very good year for autumn passage throughout Devon.

The movements of birds during the late autumn and winter depends on the weather and food availability. Most of the large autumn flocks are transient, and move out of the plantations when the food source declines. When the cone crop has been good, numbers will stay to winter, but in years of poor crops birds quickly abandon the high plantations. The largest count ever on Dartmoor was at least 400 present at Fernworthy on 19 February 1985, no doubt after a season of exceptional cone production. Other good counts during the winter period have been about 200 at Yarner Wood on 28 January 1988, the same number at Kennick Reservoir on 28 January 1994, and 140 at Fernworthy on 24 February 1997. There is thought to be correlation between the state of the cone crop and the numbers of Siskins using garden feeders in late winter. A poor crop, running short in January/February, will equate to high numbers visiting gardens and their artificial food source, and a good crop will mean fewer birds visiting human habitation (Wernham *et al* 2002). Likewise, the previous year's cone production determines when wintering birds leave to return north or to the continent, and when breeding birds return to the moor. In good years breeders may return as early as February or early March, as at Thornworthy in 1996. In poor years it may be April before they have returned. Small groups will frequently boost their food supplies by visiting garden feeders well into April if their natural food is in short supply. Several of the large flocks seen in the plantations in early spring, such as the 200 at Hennock Reservoirs on 2 April 1980, undoubtedly contain migrants coming off passage to feed before continuing their journey north.

Up until 2002 there had been seven records of ringed birds with Dartmoor connections. They show just how unpredictable the movements of individual Siskins can appear to be. The most interesting, and straightforward, was a wintering female, or possibly a passage migrant, ringed at Postbridge on 10 March 1971 and found dead at Isojoki, Vaasa, Finland on 28 April the same year. A juvenile ringed at Cwmystwyth, Dyfed on 9 July 1991, had come south for its first winter and was re-trapped at Okehampton on 19 February 1992. An adult female ringed on 25 February 1992 at Okehampton, moved west and was killed by a cat in Lostwithiel, Cornwall eighteen days later on 15 March, whilst another adult female ringed the same day at Okehampton was re-trapped at Melvich, Highland Region on 13 May 1992, having no doubt returned to its breeding site. At Dousland on 26 March 1994 an adult male was trapped and ringed, and next turned up in a net at Tongeren, Gelderland, Netherlands on 12 April 1995, perhaps having spent one winter in Britain and the next in the Netherlands.

LINNET *C. cannabina*
Resident breeder, and probable passage migrant.

The Linnet is a bird of dense scrub, particularly Gorse. As such, its habitat on Dartmoor is mostly restricted to the heathland on Downs and Commons around the moorland fringe. Pairs often nest colonially, and with small feeding flocks and pairs frequently flying off together some distance when alarmed, they can be difficult to census. The RSPB Survey of 1979 located a total of 107 pairs, with 95.7% of all territories associated with Gorse. Because of the difficulties in censusing, and patchy distribution it was thought that this total was an under-estimate (Mudge *et al* 1981). The Dartmoor Moorland Breeding Bird Survey of 2000 estimated a total of 3595 +/- 1441 birds, but here a caveat was added, suggesting that because of the difficulties in surveying, the total could be an over-estimate (Geary 2000). In a central area of the moor, the Postbridge region was found to hold surprisingly small breeding numbers in the survey from 1956 to 1967. Pairs bred at widely scattered sites, and did not exceed 10 to 25 in total (Dare and Hamilton 1968). By the early 1990s the position was still very much the same, with about 15 to 20 pairs breeding (Dare 1996). There have been very few casual estimates of breeding to supplement the above, but up to 17 pairs were thought to be present on Roborough Down in 2000, and during the fieldwork for the Devon Atlas Project 22 singing males were located in the Bonehill tetrad.

Once the young are fully fledged, family parties will roam their local patch in late summer feeding on a variety of seeds. By August and September family parties have formed flocks. Although most of these flocks are small, at least at first, by September larger congregations can be found. There is at this time a tendency for birds to move away from breeding areas. Moor edge sites can still hold flocks of a 100 or so, as with the 150 in the Cadover Bridge area on 18 September 2001, but larger numbers are to be found on agricultural land, as with the 200 at Bridford in October 1999, and the 300 there on 16 September 2000. By late October or early November the breeding population has moved off the moor almost completely, to winter at lower levels.

There follows a period in early and mid-winter when Linnet flocks are restricted to farmland, to feed in *brassica* fields, or on various weeds of cultivation. At good feeding sites flocks can reach several hundreds at this time. Examples of this were the 450 feeding in a kale field at Blackingstone Rock on 15 January 1995, and the wintering flock 400 to 600 present at Bridford from November 1999 to February 2000. As Linnets are partial migrants, it is possible that these large winter flocks could contain birds from out of the region, or even from the Continent. At Hollowcombe Head a late flock of 200 birds was present at this unusual site on 6 November 1983. This flock contained a Twite, and thus was probably made up of migrants from the north. In March flocks begin to break up, although some still exist into April. Males are by then drifting back onto territories, and nest building is usually underway by the end of April.

TWITE *C. flavirostris*
Very rare passage migrant or winter visitor.
With only one acceptable record, this Linnet of the North is a very rare bird on Dartmoor. A male was seen with a flock of Linnets, presumably migrants, at Hollowcombe Head on 6 November 1983. It is possible that others may occur with movements of Linnets in autumn or winter.

LESSER REDPOLL *C. cabaret*
Migrant breeder, passage migrant and winter visitor.
The first intimation of Redpolls breeding on Dartmoor were birds heard and seen at Fernworthy in the June of 1952. The next year two males were in song on 10 June, and on the 18 July a flock of eight or nine birds were seen, indicating that successful breeding may well have taken place. The first male was found in Soussons Plantation in 1957, and other males were found at Bellever. One to five pairs were estimated to be breeding in Soussons by 1967, but although birds were consistently seen at Bellever breeding was never proved at this time (Dare and Hamilton 1968).

On Dartmoor, Redpolls chiefly inhabit young plantations in the breeding season.

They prefer trees up to about six metres high, and when trees get more mature they are abandoned. This leads to a somewhat transient population that will be present in an area for years and then totally absent. Most of the plantations had areas of young trees in the 1960s and 1970s, and the species became widespread in small numbers. Birds were first noted at Hennock Reservoirs in 1962, and at Burrator in 1964. Summer birds were at Venford and Tor Royal by 1980 and Beardown by 1988. Breeding numbers were always small, and varied with the growth of the conifers. Away from the Postbridge area, where the numbers breeding had risen to about ten pairs by the early 1990s (Dare 1996), the highest totals were up to four territories at Burrator in 1976, and more recently three at Corndon Down in 1999. Many plantation areas were too mature for Redpolls by the 1990s.

As early as 1955 birds were being noted in the breeding season in what was apparently less suitable habitat. Mixed and deciduous woods were getting records, with birds at Yarner Wood in 1955 and 1956, and Dendles in May 1984, when a male was holding territory. It was, however, the bird's more traditional habitat of alder and willow scrub that attracted most breeding pairs, away from the young plantations. Birds were noted prospecting willow scrub in the West Webburn Valley in 1963, and by 1967 three pairs were on territory at Challacombe. In 1969 at least four pairs were present in similar habitat on Prewley Common. More recently birds have been noted in willow scrub habitat at Vitifer and Bagtor Newtake, where a party of at least 10 including juveniles, were seen on 12 July 2002.

Birds disperse outside the breeding areas from August on, and become scarce. The maximum counts for autumn have been only seven at Fernworthy on 11 September 1963, and nine in flight over Huccaby on 25 August and 29 October 2001. By November numbers begin to pick up, and small flocks are noted in riverside alders and also in the plantations and other woodland. The biggest flocks during winter have occurred at Yarner Wood, where early January quite regularly sees large numbers. The first week of January in at least eight years since 1979 has produced numbers exceeding 50, with a huge 150 being present on 24 January 1988. Elsewhere numbers are far more erratic, but 35 at Hennock on 1 January 1980, 26 at Throwleigh on 28 March 1990, 30 at Portworthy on 27 December 2000, and 25 in birches on the Dewerstone 28 February to 1 March 2002 are worthy of note. A rather strange movement of birds over open moor was observed on 3 March 1996. In the morning six birds flew north up the Yealm near Yealm Falls, and later in the day a further 10 flew east, calling, over Penn Beacon. The winter of 1995/1996 had been an especially good one in Britain for wintering Redpolls, and presumably this movement consisted of migrants moving back east.

British breeding Redpolls are considered to be partial migrants. The wintering area for the small Dartmoor population is unknown. They may stay in the lowland areas of the county during the cold months, taking advantage of birch catkins or riverside alders, or they may move further, perhaps to the continent. The flocks that arrive in the area, especially in late winter, no doubt contain migrants from further north, if not Northern Europe, but this has not been proven. No birds of the Scandinavian and North Russian nominate race Mealy Redpoll *C. flammea,* or the related Arctic Redpoll *C. hornemanni* have ever been recorded in winter flocks on Dartmoor, although birds of the former could occur in feeding parties at this time of year. By April wintering and passage flocks have dispersed or moved on, and local breeders have returned to take up territories.

Crossbill

CROSSBILL *Loxia curvirostra*
Irruptive passage migrant and winter visitor, which sometimes stays to breed.
A loud chorus of "chipping" calls overhead and brief glimpses of large-bodied finches skimming the conifer tops in flight is the first indication that most observers get that a Crossbill irruption has taken place. The irruptive movements of Crossbills are triggered when, in their Northern European breeding areas, a good breeding season is followed by a poor cone seed crop. Being almost totally reliant on conifer seeds for food, birds have to move away from their nesting sites at once. They may only have to move a short distance to locate a further supply, but if the crop is poor over a wide area they have to move huge distances, and a major exodus takes place.

221

Crossbills are very early breeders, with birds breeding from November to April, when spruce seed is most available (Wernham *et al* 2002). After this period they frequently have to move to locate new spruce crops. Birds arrive on the east coast from the continent in late May, but it is usually June, or later, before any migrants are seen in the Dartmoor plantations. Numbers can vary greatly, from a very few most years, to a huge influx occasionally, depending on the weather during the time of migration and the size and scale of the irruption. Another factor of course, is the extent of our local conifer crop. If it is a bad year, birds may arrive, but pass on very quickly almost unnoticed. In a good year, they may stay in numbers, and be present for many months. Some then become resident in the plantations and breed the following year. Sometimes these immigrants build up a small local population, but they tend to disappear after a few years and leave the plantations untenanted until the next large irruption.

Arrivals of Crossbills have been noted in Britain for centuries. D'Urban and Mathew mention invasions in June 1837 and July 1868, and birds killed frequently in the neighbourhood of Ashburton, and at South Brent in 1889. Large invasions that come as far west as Dartmoor are rather infrequent, and some of the earlier influxes into Devon were only poorly recorded on the moor. In more recent years birds have been seen in numbers in 1966, 1972, 1990, 1993 and especially in 1997 when the biggest invasion ever was recorded. The usual pattern is for birds to arrive from late June, and for numbers to build up during July and August, if the cone crop is sufficient to hold them. There is then, at times, a further wave of birds in October or even November. Sometimes, as in 1996, this October wave is the only arrival that occurs. Birds normally feed on the seeds from spruce cones, but larch cones also prove very attractive to them, and indeed in February and March 1984 many hundreds were feeding in the larch plantations at Hennock Reservoirs. Beech mast is also very acceptable and during October to December 1990 at Fernworthy 30 to 35 birds regularly fed on the massive beech mast crop. In invasion years flocks in most plantations are usually in the region of 20 to 50 birds, but at times this is well surpassed. Hennock plantations have proved the most attractive for Crossbills with 100 estimated there in July 1997 and September 2002, over 150 in March 1980 and March 1994, and over 200 in March 1982 and January 1994. Fernworthy held 90 in November 1996, and between 120 and 200 in October/December 1993. Soussons attracted about 100 birds in January 1994 and July 1997, and Burrator had an estimated 200 in June 1997 and 100 the following month. Birds can be seen crossing the open moor between plantations at times, as with the five that flew north, low, over Shavercombe Head on 29 June 1953, the 40 well out on the moor near Wrangaton on 21 September 1966, and the five going north at Leedon Tor on 27 August 2001.

By the end of March the flocks have dispersed, and most birds have started back for northern forests. The pairs that stay are well into the breeding cycle by this time, and in fact may have young fledged. The earliest fledged juvenile on Dartmoor was one being fed at Fernworthy on 8 February 1981. Successful breeding is not easy to confirm, and it is probable that more pairs breed than are ever located. The first breeding was at Soussons in 1975, when a pair with two juveniles was watched on 8 June. Since then, confirmation has only been obtained half a dozen or so times, although it has been suspected a great many more. The best year would appear to have been 1997, when after a late influx of birds in December 1996, pairs bred at Fernworthy, Soussons and Burrator, probably at Bellever, and quite possibly at other locations, assisted by the excellent cone crop.

BULLFINCH *Pyrrhula pyrrhula*
Resident breeder.
The Bullfinch is a species that is regularly seen, be it usually only fleetingly, but little understood. It inhabits scrub and thick foliage in woods and hedgerows around the moor edge, and at suitable places in the interior, but although well distributed, seldom seems common. It leads a secretive life and seldom appears far from cover.

Between 1956 and 1967 the population of the Postbridge area was estimated to be 10 to 20 pairs. Numbers were thought to be increasing as the conifer plantations were colonised. About half the pairs bred in Bellever plantation, with the rest in Soussons, the garden shrubberies in Postbridge, and the deciduous woodland and scrub in the West Webburn Valley (Dare and Hamilton 1968). By the early 1990s 10 to 20 pairs were thought to be breeding in the plantations, plus another eight in areas around Postbridge and the West Webburn Valley (Dare 1996). A single pair bred intermittently at Harford Common Bird Census site between 1992 and 2000, but the more casual records do very little to expand our knowledge of this bird on Dartmoor. Sightings are widespread, but birds are seldom seen in numbers higher than a family party, and there are no indications of migrants, as there are on some coastal locations in late autumn. The population therefore appears to be sedentary and the only difference in population is after an exceptional breeding season, when numbers are noted to have increased over large areas. Such a year was 1996, when several observers noted they were numerous in June and July. Unusually large groups have been 13 at Moorhaven, Bittaford on 14 December 1986, 12 at Fernworthy on 27 October 1989, and 10 at Trendlebere on 19 January 1988. The highest count, however, came from Fernworthy on 7 August 1948, when a total of at least 30 were seen, including 12 in one flock. This was, presumably, the result of a particularly productive breeding season, rather than an indication that migrants were present.

HAWFINCH *Coccothraustes coccothraustes*
Rare winter visitor, and passage migrant.
There have been about 25 records of Hawfinches on Dartmoor up until the end of 2002. Most have been single birds, occasionally two have been seen, once three, and on one occasion a small flock was present. Birds have been seen in every month of the year except July and September. There is no clear pattern to the occurrences, but some

have been in the late autumn October/early November period, when a light passage is sometimes detected on the coast, and several others have been seen in March and April, when a return spring movement is at times noted at coastal migration points (Wernham *et al* 2002).

Birds have been seen in several wooded areas around the moorland fringe, with Fernworthy being the most regular spot. The first bird here was on 18 November 1988, but within days seven were present feeding on beech mast. They were only seen on 21 and 22 November, but two were still present on 26 November. 1988 was an exceptional year for the species in Devon. Just to the east of Dartmoor, a flock was building up at this time at Doddiscombsleigh. By mid-December this gathering totalled 44 birds, and probably included the Fernworthy seven. Two were present at Fernworthy on 20 January 1989 and 23 November 1996, with singles there 5 March 1989, 4 December 1996, 16 January 1997 and 15 February 1997. The southwest corner of the moor has had records, with two at Burrator on 8 February 1970, and singles there 15 August 1977 and 15 and 16 April 1979. Nearby a bird fed in a kale field with a finch flock on 27 January 1979 near Brisworthy Plantation. The wooded valley of the River Walkham has had three records, with one at an unspecified locality on 2 April 1938, one at Horrabridge on 13 March 1962, and one at Dittisham on 18 October 1994. Also in this corner of the moor were singles at Yelverton on 2 April 1931 and 27 October 1954.

The area around Okehampton has had its records too. Two were seen on 2 November 1988, and one at Rectory Wood on 18 August 1995. On the northeast side of the moor a bird was near Castle Drogo on 28 December 2001, and slightly further south one was at Chagford on 30 December 1947, with three together there on 7 March 1989. A single bird was seen at Christow on 26 March 1948, and on the southern edge a bird was watched feeding on haws with thrushes at Lud Gate, Buckfastleigh on 25 November 1990.

Three records concerned birds during the late spring/early summer in possible breeding habitat. One was seen in mixed beech and oak wood at Holne Chase on 6 June 1941, a female was watched in oaks near Sourton Down on 21 and 22 May 1993, and one was near Ivybridge on 13 May 1999. There has never been a confirmed breeding record for Dartmoor, but recently there have been records of birds breeding in valleys not far outside the eastern boundary of the National Park.

BLACK-AND-WHITE WARBLER *Mniotilta varia*
Very rare vagrant
On 3 March 1978 at Whitchurch, Tavistock, Iris McEwen was standing looking out of the kitchen window whilst making a cup of tea. A bird flew onto her storeroom roof and, thinking it was a Pied Wagtail, she stayed still to watch it. She quickly realised her mistake. The bird moved closer with a movement that was a cross between that of a mouse and a Nuthatch. The markings on the bird's head resembled a miniature Badger, with very sharp black and white stripes. The rest of the body was also striped

black and white. The stripes ran lengthwise along the body and wings. It was about the size of a sparrow, and had a fairly short tail. After approaching to about two feet, it flew off into nearby conifers. After consulting several books, it was identified as a Black-and-White Warbler (McEwen 1981).

This was only the fourth record in Britain of this North American wood warbler, the first for Devon, and of course the first for Dartmoor.

LAPLAND BUNTING *Calcarius lapponicus*
Very rare winter visitor
The inclusion of this species rests on just one record. At the end of January 1934 one was seen near Chagford, probably at Thornworthy, and accurately described. It was accepted in the *Annual Report of the Devon Bird Watching and Preservation Society* for that year, but a later county avifauna (Moore 1969) rejected the record. As the species gets reported rather more now than in Moore's day, and as the finder of the bird was a gentleman who still resides in the area and has studied birds all his life, it does not seem unreasonable that this record should be accepted.

SNOW BUNTING *Plectrophenax nivalis*
Rare passage migrant and winter visitor
In the late nineteenth century Pidsley stated that a Mr. Mitchell had only once met with a Snow Bunting in Devon and that was a solitary wanderer, observed on one of the highest tors of Dartmoor in the middle of winter. D'Urban and Matthew had never encountered them on the moor but thought them not infrequent in autumn and winter, and stated that they had been seen on Ashburton Down in successive winters.

Although the term "not infrequent" would now be thought much too optimistic to describe their status, birds still turn up largely in the same habitat as mentioned in the nineteenth century accounts. Moor edge habitat at Belstone, Gidleigh, Wrangaton, Fernworthy, Okehampton and Bridestowe have had records over the past 40 years or so, but most have been reported in autumn and winter from the high moor, where they feed on grass and heather seed in short vegetation and beside paths etc. The sites that have had birds most regularly are Okement Hill, the area around Yes Tor, Fordsland Ledge and West Mill Tor, Cut Hill and Hangingstone Hill. Whitehorse Hill and Rattlebrook Hill have also had reports in this northern part of the moor. Further to the southeast the whale-back of Hamel Down has attracted birds twice, and there have been single reports from Sittaford Tor, Birch Tor and Corndon Down. On the southern moor, the slopes of Holne Moor, Ryders Hill and Huntingdon Warren have all had single migrants, whilst on the boundary with the "in country" the grass-covered incline of Butterdon Hill had a wintering bird in 1947. Two were around the old clay works at Redlake in November 1999.

The peak autumn migration time on the coast is also the time when most have occurred on Dartmoor. There have been 17 records in this period, from the third week of October, until the middle of November. There have been three records in early October, and two in the last week of September. The earliest was a bird at Bellever on 27 August 1998. All birds have been one-day sightings, and no doubt had come off passage to feed before moving on. Indeed, a male on the summit of Hamel Down on 22 October 1968 was seen to arrive from the north and fly on south along the ridge after feeding briefly (Dare 1996).

Wintering birds have not been so well reported as migrants, but this is not surprising when one considers that many fewer people are walking the high moor in the depths of winter, and nearly all of the winter birds have been seen on the wind-swept tops. There have been 12 records from late November to the end of February. There has been an unconfirmed report of a small flock of birds present in a recent winter in the Yes Tor/High Willhays area, so perhaps there is a small regular wintering population that has yet to be recorded. March has had two records of birds which could have been returning migrants, but the only undoubted spring migrant was one at an unpublished Dartmoor location on 14 April 1934.

YELLOWHAMMER *Emberiza citrinella*
Resident breeder and possible passage migrant.

The Yellowhammer has always been a well known, and perhaps somewhat taken for granted, bird of the British countryside. Its familiar " little - bit - of - bread - and - no - cheese" song has been one of the regular sounds of spring to countrymen for generations. However, in the last 10 years or so there has been a national decline in numbers. This decline, as with other Bunting species, has been due to some extent to loss of habitat, but mainly to changes in agricultural practices causing a lack of sufficient seed food in winter.

The main habitat is the moorland fringe areas, with high densities occurring in areas of gorse scrub and bracken, often on the edges of moor and agricultural land. In the 1950s and 1960s a particularly dense population was found in the Postbridge area, where 50 to 100 pairs were estimated to be present. This was due largely to the numbers inhabiting the young conifer plantations (Dare and Hamilton 1968). When these trees matured, the habitat was no longer suitable, and numbers reduced. By the early 1990s, only nine singing males could be found (Dare 1996). The 1974 Breeding Status Survey organised by the Devon Bird Watching and Preservation Society found that although birds were well distributed generally, there were areas of suitable habitat where they were difficult to locate (Sitters 1975). Several of these areas were around the edge of Dartmoor. In the 1979 DNPA/RSPB survey of breeding birds on Dartmoor, the Yellowhammer was found to be a common although locally distributed species. A total of 153 pairs were located at lower altitudes with none over 1300 feet. Breeding habitat was largely confined to moorland edges, with a gap in distribution in the north and northwest (Mudge *et al* 1981). The next survey to estimate numbers was the 2000 Dartmoor Moorland Breeding Bird Survey for the DNPA. By this time the species nationally was in a state of decline, with 43% reduction between 1970 and 1998. On Dartmoor, however, numbers appeared to be holding up well. They occurred at a high density in gorse and bracken dominated habitats and the estimated overall population was 526 +/- 259 pairs (Geary 2000). There was a caveat however, that because of the survey methods and areas covered, the population could have been over-estimated.

Elsewhere there have been few records that give any insight into the rise or fall in breeding numbers at localities over a period of years. One noted exception has been the Harford Common Bird Census site, where two territories during 1988 to 1990 had risen to five in 1991, only to decline to two or three again up to 1996, after which no birds bred. Also very useful are the figures from Dunsford Wood Nature Reserve on the eastern borders of the National Park. Here the figures have been steady from 1980. Five territories were usual from 1980 up until at least 1993, increasing to six in 1992 only. Four were present in 2002. A good example of an area with a high density colony-type population is Roborough Down. Here, especially in the southern portion of the Down, a habitat of gorse, bracken and occasional hawthorn and rowan trees held up to 40 pairs/singing males in 2000, although it must be added that by 2004 numbers appeared to have reduced somewhat. So conclusions are mixed as to the present status of this species around the moor, some areas seem able to sustain a good population, whilst others have very few. It is certainly a breeding bird that should be monitored closely in the future.

Yellowhammers are largely sedentary on Dartmoor, so the small flocks that are seen in autumn usually consist of a few local family parties. The largest of these would appear to be the 20, mainly juveniles, seen at Bridford in August and September 2001. As autumn turns into winter there are signs of an altitudinal movement. At Wootton, Holne in 2002 birds were noted flying over on eight dates between 4 November and 21 December. Most were singles, with various flight directions recorded. These were most likely birds dispersing from moor edge territories to lower ground, although migrants from elsewhere cannot be ruled out. A flock of 12 flew west-northwest over Holne on 18 November 2003. By December most have moved off the high moor. However, birds can be enticed to stay when tempted by feed put out for them. At Huccaby in 2001 seven stayed into December, taking advantage of the fields that had become rank and seed rich during the summer, when they could not be touched due to the foot and mouth crisis, and also because seed was put down for them. None had been present in previous winters. Another instance of birds taking advantage of an available food source and not moving far off the moor was the presence of at least 100 loitering around Pheasant feeders at Gallant le Bower on 21 January 2004. This was an astonishing number to be on Dartmoor in winter, equalled only by a flock of about the same size in the much more agricultural surroundings of Dunsford in late December 2001. Birds will regularly visit gardens during the winter if food is provided, as with up to 15 at Yelverton in the 2002/2003 and 2003/2004 winters. By February, if the weather is sufficiently mild, males are beginning to take up territories again, and the familiar song is often heard again from mid-month. Flocks can still be found, however, into spring, and it is probably well into April before all territories are occupied.

CIRL BUNTING *E. cirius*
Very rare resident breeder.

The Cirl Bunting is on the very northern edge of its range in southern Britain. After disappearing from many regions in recent decades, its last remaining refuge has been along the south coast of Devon, particularly the South Hams, where, with changes in certain agricultural practices, it has made a comeback from a very low base level in the last 10 years.

On Dartmoor, there is no indication that it has ever been anything other than a very rare occasional breeder. Over the years birds have occasionally moved north from their lowland coastal haunts to take up territories around the moor-edge, but reports have always been very sparse. The valley of the Dart has attracted birds at times. There is a statement in the 1948 *Annual Report of the Devon Bird Watching and Preservation Society* to the effect that no birds had been seen at Buckfastleigh that year, and this was very unusual. They were noted back in the area again, however, by 1951. A bird was seen at Holne on 3 August 1930, and in 1994 there were records of birds in fields around Stokes Farms, Holne, on the edge of the moor. Further upstream on the East Dart a male was singing at Sherwell Farm, north of Dartmeet on 20 May and 1 July 1979. This is the nearest to the centre of the moor that the species has been recorded.

Other records have tended to be around the periphery, with a male in a garden near Cosdon Beacon on 3 June 1933, one at Wrangaton, seen and heard on 7 March 1941, one seen on several occasions at Yelverton over a two month period from 26 March 1946, and a single at Chagford on 19 May 1964. More recently breeding pairs or singing males have been recorded in territory in two areas on the southern edge of our area. One or two males were present in horse paddocks on the fringe of agricultural land at Haytor in June 1994, and again in May 1998, and two or three pairs have been resident around the Teign Valley Golf Course since 1995. One was in a garden at Ivybridge on 3 March 1999, and at the eastern edge of the area, a pair bred at Dunsford in 2002. The most recent record concerns a singing male in the china clay country at Wotter Pit on 7 May and 17 June 2004.

RUSTIC BUNTING *E. rustica*
Very rare vagrant.

On 22 and 23 March 1997 a Rustic Bunting joined the familiar garden birds around feeders at a house in Dousland. It was the first for Dartmoor and only the sixth Devon record. All other records had been of birds on Lundy during the month of October. This was also the first of three rare bunting records in spring within a few years in this southwestern corner of Dartmoor. The individuals were no doubt returning eastwards after being disorientated on passage the previous autumn, but where they had wintered, and why they should turn up inland within a few miles of each other, is an intriguing mystery.

LITTLE BUNTING *E. pusilla*
Very rare vagrant.

This, the second of the rare spring bunting records, was found in a cottage garden near Kelly College, Tavistock on 16 March 1999. It fed with the regulars, and appeared intermittently until 7 April. What made this record even more amazing was the fact that the bird returned in the autumn and spent some time in the garden between the 17 and 21 November. The garden was in fact just outside the National Park boundary, but has been listed here for completeness. It was the first record for the Dartmoor area, and the eleventh for Devon.

REED BUNTING *E. schoeniclus*
Resident breeder, and possible passage migrant.

In the late nineteenth century, neither D'Urban and Mathew nor Pidsley gave any indication that the Reed Bunting had ever been regular on Dartmoor's bogs and mires. Indeed the former only mention it as being found in the meadows near Ashburton, and the latter states that it is very seldom seen on Dartmoor. In the 1930s and 1940s the *Annual Reports of the Devon Bird Watching and Preservation Society* still included every record and breeding pair for the high moor. It was not until the survey work in the Postbridge area between 1956 and 1967 that the status of the species was found to be other than a rather scarce, irregular breeder. To be fair, this may be an over simplified inference, as Postbridge had a new emerging habitat at that time in the form of the young conifer plantations, which proved ideal for Reed Buntings. However, an estimated population of 25 to 50 pairs in the bogs and plantations (Dare and Hamilton 1968) was hugely in excess of the handful of records from elsewhere on the moor. By the early 1990s, the plantation habitat had been long lost due to the maturing of the trees, and the population had dropped to about 15 to 20 pairs (Dare 1996).

Casual records during the 1970s were still rather sparse, but the DNPA/RSPB Survey of 1979 found the species reasonably abundant and widespread. The 68 pairs located during the survey were found in mostly wet areas from the moor edge to the margins of the high plateau areas. Nesting habitat was characterised by the presence of Soft Rush, which was found in 90.5% of territories. Only three of the territories were in dry habitat; two were on Bracken slopes, and one on a grass slope with some mature Heather (Mudge *et al* 1981). The acceptance of habitat away from water was noted in several areas of Britain in the late 1960s and 1970s, but instances on Dartmoor appear to be few. A pair nested in Heather above Burrator in 1937, and a singing male was seen and heard in a roadside hedge below Buckland Common in 1975. Territories on the largely dry Roborough Down were found in 1951 and 1974, and a male was singing from a thorn bush on the upper slopes of Birch Tor in 1997. The most recent record is of a male singing, again from a thorn bush, in the typical Yellowhammer habitat of High House Waste in 2004.

Between 1979 and the next full survey in 2000, Reed Buntings became more numerous on the moor. Casual records had been increasing steadily, and it was obvious that a range expansion had taken place, with pairs inhabiting nearly all marsh habitat, even very small areas, right across the high areas including the central plateau. The Dartmoor Moorland Breeding Bird Survey of 2000 estimated the total population to be 601 +/- 189 pairs, and to be of regional importance. This survey covered the Rhôs pasture systems around the moor, which was not covered in 1979, but even allowing for this and acknowledging that the total in 1979 was probably underestimated, there had still been a large increase. The highest breeding densities were found in the Rhôs pasture systems and in moorland valley mire (Geary 2000). The extent of this increase is rather surprising for two reasons. Firstly, it is against the national trend, which had shown a decrease of 64% in the population between 1972 and 1996 (Mead 2000), and secondly birds are known not to favour upland areas in the north and west of Britain (Wernham *et al* 2002, Sitters 1988), and the New Atlas 1988-1991 showed a contraction of range in the South-West (Gibbons *et al* 1993). Fieldwork for the Devon Atlas Project in 2002 showed a population that was still healthy, with good totals including nine territories in the Childe's Tomb tetrad, four territories in the Bonehill tetrad, three pairs at Ditsworthy Warren, and at least 10 males along the Swincombe from the Intake Works to Wydemeet.

In the autumn family parties disperse from their immediate breeding area. Most do not at first go far, and birds can be found searching areas of Bracken and other cover that they would not normally be associated with in the breeding season. It is possible that in the late autumn migrants, from further north in Britain or perhaps the continent, pass through our area. With the coming of the colder weather in November/December most birds leave the open moor to take advantage of milder conditions at lower altitudes. How far they go is not known, but it is probable that most stay within the county. Some congregate at suitable moor edge sites. Fernworthy has a regular small flock in the winter, the recent maximum being 24 there on 27 January 1998. Another moor edge site is Yarner Wood, where about 40 were found on heathland on 15 January 1989. Others will stay on the moor, providing a suitable feeding site is found. Twelve in a kale field near Runnage on 4 January 2004 were an example of this, and even the bleakest locations can hold birds at times, like the individual found in rushes near Avon Head in severe weather on 29 December 1996. Two males wintered in a garden at Okehampton in 1979.

By mid-March the first males are back on territory, with most returned by early April.

BLACK-HEADED BUNTING *E. melanocephala*
Very rare vagrant.
The last in the trio of rare buntings was a male of this species that frequented a private garden at Peter Tavy from the 20 to the 26 May 2000. The date was typical of spring arrivals for this species, but the locality certainly was not.

CORN BUNTING *Miliaria calandra*
Very rare breeder, and possibly passage migrant/winter visitor.

This is a species of cultivated farmland that shuns high ground and wooded areas. It has always been coastal in Devon, and its distribution very patchy, making it a very scarce to rare resident in the county. It is thus strange that Dartmoor has records at all, but a pair bred at Lydford in 1971 and 1972, and a bird, with possibly two others, was seen at Horrabridge on 18 November 1983.

As with many bunting species, the Corn Bunting has declined dramatically in many areas of Britain in recent years. It has all but disappeared from the Southwest, only just holding on in a narrow coastal strip on the north Cornish coast, and there is little chance that it will ever breed again on Dartmoor.

SPECIES INDEX

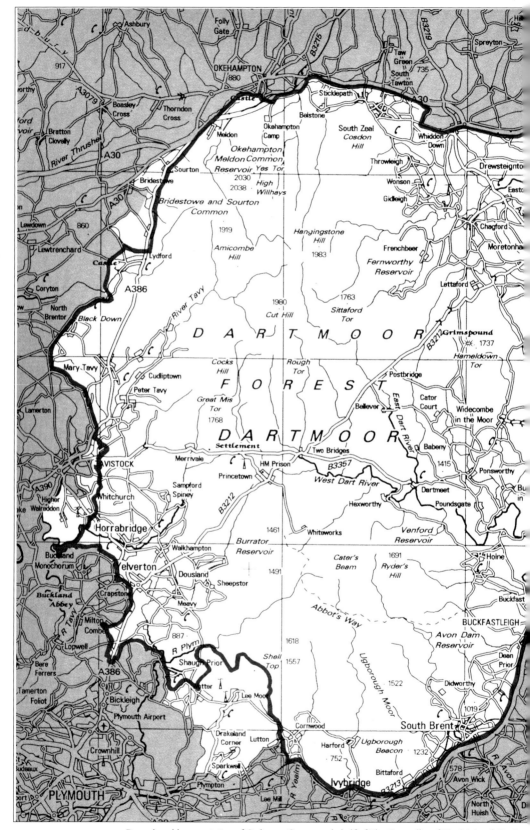